Samah KARIM
Abdelouahad AOUNITI

Characteristics of wastewater

Samah KARIM
Abdelouahad AOUNITI

Characteristics of wastewater

of the cooperative Laitier du Maroc Oriental (COLAIMO)

ScienciaScripts

Imprint

Any brand names and product names mentioned in this book are subject to trademark, brand or patent protection and are trademarks or registered trademarks of their respective holders. The use of brand names, product names, common names, trade names, product descriptions etc. even without a particular marking in this work is in no way to be construed to mean that such names may be regarded as unrestricted in respect of trademark and brand protection legislation and could thus be used by anyone.

Cover image: www.ingimage.com

This book is a translation from the original published under ISBN 978-620-2-27795-2.

Publisher:
Sciencia Scripts
is a trademark of
Dodo Books Indian Ocean Ltd. and OmniScriptum S.R.L publishing group

120 High Road, East Finchley, London, N2 9ED, United Kingdom
Str. Armeneasca 28/1, office 1, Chisinau MD-2012, Republic of Moldova, Europe
Printed at: see last page
ISBN: 978-620-5-76755-9

Contents

Acknowledgements 2

Resume 3

Introduction..4

Part I..5

Part II...23

Part III..33

Part IV 40

Part V 48

Conclusion 61

Bibliographic references 62

Acknowledgements

Before elaborating on this professional experience, it seems appropriate to begin this report with a word of thanks.

First of all, I would like to express my sincere thanks and deep gratitude to my pedagogical supervisor Mr. Abdelouahad AOUNITI, for his efficient supervision, his encouragement, his pedagogical work, his indefectible help, and for his great availability. My gratitude also goes to the members of the jury Mr.B.HAMMOUTI, Mr.A.DAFALI and Mr.C.BELBACHIR who accepted to judge this work.

I would like to thank Mr TAHIRI Hassan, director of COLAIMO, who gave me this technical and enriching training course. I would also like to thank Mr Mhamed AMRANI, my supervisor at COLAIMO, for all his precise explanations.

I would like to thank Mr. Kamal BEKKOUCH, coordinator of the Master's degree in Water Chemistry, as well as all the faculty members of the Faculty of Sciences of Oujda for their continuous efforts to propel our training into the realm of excellence.

I would like to express my deep gratitude and sincere thanks to Mr. CHAOUKI BELBACHIR, and Mr. Mohamed SBAA who helped me a lot to carry out the analysis of my samples.

I would like to thank all the engineers, technicians, agents and workers of COLAIMO who have accompanied me throughout this experience with a lot of patience and pedagogy.

Finally, I would like to thank all the people who have contributed to the realisation of this work and to make this course a very profitable moment.

Resume

The present work focuses on the characterisation of wastewater from the Cooperative Laitiere du
The aim of this study is to identify the main sources of water in the Eastern Morocco region

(COLAIMO) and to recommend an appropriate treatment to allow their reuse, thus reducing

the nuisance to the receiving environment and also to remedy the loss of this water source in

recoverable materials.

The physico-chemical characterization of the raw wastewater revealed that this liquid

discharge is very

organic matter load in terms of COD which is 3900mg O_2 /l, and 2300 mgO_2 /l of
BOD5. Furthermore, the COD/BOD5 ratio = 1.69 underlines the biodegradable character of

the COLAIMO wastewater, for which a biological treatment seems to be quite suitable.

Biological treatment processes are diverse but after a comparative study of the different

processes and a technical study evaluating the requirements of the cooperative, the

"Sequential Biological Reactor (SBR)" process appears to be the most suitable for this type of

industry and is the most efficient.

For the implementation of a wastewater treatment plant, a design study of the plant prior to its

implementation is very interesting. The objective of this report is to carry out a design study

of the SBR treatment plant chosen for the treatment of COLAIMO's dairy effluent for a

sustainable management and operation of the planned plant.

Keywords: dairy effluent, COLAIMO, purification, Sequential Biological Reactor, design.

Introduction

In the training programme of the Master specialised in water chemistry, an end-of-study internship is compulsory, aiming to consolidate our knowledge by confronting the reality of the field, to discover the world of business and to familiarise ourselves with professional life.

It is within this framework that I carried out an internship at the Eastern Moroccan dairy cooperative (COLAIMO), which plays a key role in supplying the market in this region with milk and its derivatives.

At the beginning of this internship, I made a general tour in the different departments of COLAIMO in order to discover the manufacturing and packaging processes of milk and its derivatives, focusing mainly on the water circuit from the plant supply to the wastewater generated during these processes to have an idea on the origin of this wastewater as well as the polluting load of the plant.

The present study focuses on the characterisation of the liquid effluents of COLAIMO, in order to evaluate the pollutant load of the discharges and to identify the appropriate treatment process, with the aim of valorising this wastewater and thus fighting pollution and preserving water resources in the Oriental region.

This work is carried out thanks to a series of measurements carried out on samples taken from the main collector of the cooperative's liquid discharges. The physico-chemical and microbiological analyses were carried out in the laboratory of the Oriental Centre for Water Science and Technology and the Regional Analysis and Research Laboratory of ONSSA.

This paper is composed of five parts: the first part is devoted to a bibliographical study showing the state of the problem of the industrial liquid discharges, by giving the various techniques of purification of the liquid discharges as well as the financial opportunities opened for this type of investment, the second part includes a description of the dairy cooperative of Eastern Morocco. ^{eme}The third and fourth parts present all the materials and analytical methods used in the framework of this study of the characterization of liquid waste, as well as the results obtained during this study and their interpretation. The last part is devoted to the study of the design of the planned treatment plant and the different proposals for the valorisation of these dairy effluents.

Part I
Bibliographic synthesis

I. Water, the strategic issue of tomorrow :
1. <u>Water resources, demands and deficit:</u>

Morocco is faced with a water shortage due to a drought that has increased over the past two decades as a result of population growth, industrial and domestic consumption and the overexploitation of groundwater, particularly in agricultural areas. If Morocco's growth rate is maintained, consumption will have to go from 830 m3/ha/year in 1990 to 411 m3/ha/year in 2020, i.e. half the amount, while more than 70% of the potential is mobilised (Beauchamp, 2003). In 2009, the cost of damages related to water degradation represented more than 1.23% of the national Gross Domestic Product, i.e. more than 33.15% of the country's total estimated environmental damages according to the National Environment Council (Conseil National de l'Environnement, 2009).

Faced with a demand for water that is constantly increasing in the Oriental region, known for its semi-arid climate, supply remains limited due, on the one hand, to the reduction in the volumes mobilised by the various hydraulic works and the exhaustion of groundwater by uncontrolled pumping and, on the other hand, to the deterioration in the quality of these resources.

Water needs will increase even more in the future due to the intensification of agriculture, urban development due to an increase in the population, especially in urban areas, the creation of new water-consuming tourist projects such as the large seaside resort of Saidia, and finally to the industrial development planned in the future with the creation of a technopole in Oujda, an agropolis in Berkane and new industrial zones in the region.

Indeed, all the projections made in the short term show that these future developments will inevitably have two major consequences whose impact on the development of the region may prove fatal and should therefore be taken into account in any planning. On the one hand, the excessive exploitation of the resource would lead to a quantitative imbalance between supply and demand and the establishment of a structural water deficit which would be difficult to restore; on the other hand, the activities resulting from these developments would affect the quality of the resource.

Water will therefore be one of the main challenges in the coming years. While waiting for the implementation decrees of the water law 10-95, industrialists must introduce the concept of clean technologies in their overall management. Pollution must be treated not only downstream of a production unit (at the end of the pipes) but also at the point of pollution generation. Recycling and reuse of liquid effluents is one of the bases of this concept. The example of a European dairy is illustrative in the field of water management.

Indeed, the De La Vallee De L'Ourcq dairy in Rungis, France, discharges 500 m3 of liquid effluent daily, carrying 1.5 tonnes of BOD5. After the implementation of clean technologies, more than 350 m3 of water are recycled. Compared to the initial state, the dairy has reduced its discharges by 70% and consequently reduced its withdrawals from the natural environment by the same percentage. In terms of pollution, the flows of suspended solids are reduced by 100% and those of BOD5 and COD are reduced by 90% (Castillo, 2005).

As with any industry, water consumption in the processing sector is not subject to any legal text. It is the industrial liquid discharges that are under the legal sway of the water law 10-95 and its application texts. This law reinforces the "polluter pays" principle and defines the limits of water pollution by industrial sector. At present, there is an operational "Norms and Standards" committee which brings together all the bodies concerned. This committee capitalises on the achievements in the field of standardisation and is active in the process of developing national standards.

The dissemination of good environmental practices in the industrial sector must be preceded by preliminary fieldwork to assess the experience, willingness and understanding of the main industrialists with regard to water management in their plants; to identify opportunities for water savings and reduction of pollution loads; and to identify the technical and financial mechanisms to transform these opportunities into feasible projects.

2. The problem of industrial wastewater :

Industrial wastewater refers to water that originates from industrial activities. Industrial wastewater is different from domestic wastewater and its characteristics vary from one industry to another. In addition to organic, nitrogenous or phosphorous materials, it may also contain toxic products, solvents, heavy metals, organic micropollutants and hydrocarbons. Some of them must be pre-treated by industrialists before being discharged into the collection system. They are only mixed with domestic water when they no longer present a danger to the collection networks and do not disturb the operation of the treatment plants.

> Industrial pollution :

The industrial pollution load discharged into the Moulouya system is currently estimated at around 2,500 tonnes of oxidisable materials. This load will increase to about 3,600 tonnes by 2020 if no clean-up scenario is put in place by the stakeholders concerned. Currently, 85% of the industrial pollution load, i.e. 2,100 tonnes of organic matter, is discharged in a diffuse manner directly into the natural environment (wadi and soil) and only 15% of the industrial pollution, i.e. 368 tonnes of organic matter, is collected by the city sewerage system.

Table 1: Industrial pollution parameters of different cities in l'oriental (A.B.H.M ,2010)

Province	Flow rate m3/year	BOD kg/year	COD kg/year	TSS kg/year	OM kg/year	PT kg/year	NTK kg/year
Nador	384 400	417 200	631 850	244 410	488 750	910	30 400
Oujda	108 600	296 350	430 020	62 340	340 910	2 500	22 940
Taourirt	37 950	18 700	33 060	10 910	23 490	520	4 990
Taza	33 470	1 006 100	2 771 610	32 420	1 594 600	9 770	6 840
Berkane	1 970	250	1 620	160	700	-	49 200
Total	566 390	1 738 600	3 868 160	350 240	2 448 450	3 700	114 370

The rate of industrial depollution recorded in the Moulouya basin remains relatively low and does not exceed 19% of the discharged load (ABHM, 2010). This rate shows the lack of awareness of some industrialists to clean technologies. Efforts to clean up pollution must still be pursued at the basin level, particularly for isolated polluting units that are not connected to the sewage system, in order to comply with direct discharge standards and protect water resources.

In fact, pollution control measures are only carried out at the level of large structured companies such as SUCRAFOR, SONASID, ATLAS BOTLLING, the Tannery of Bni Snassen or in the units working as subcontractors for large international brands and which are required to respect certain international standards. However, despite the fact that these plants have wastewater treatment facilities, their discharges are sometimes loaded with pollutants and micropollutants. In this case, efforts to monitor the quality of the discharges and to clean them up must be continued in order to comply with the discharge standards. Similarly, at the level of traditional olive crushing plants, pollution problems are far from being a priority. These units are very numerous and discharge 65% of the total organic load of the basin, especially in the province of TAZA, Guercif and Taddart. The margines generated by these units are either stored in very leaking basins ensuring their infiltration into the subsoil, or collected by tanker trucks and then discharged without any treatment into the natural environment. These practices can cause significant risks of contamination of groundwater resources.

II. General environmental aspects of the dairy industry

The main environmental aspects of the dairy industry are related to high water and energy consumption, the generation of wastewater with a high organic content, and the production of waste. Finally, emissions of gases and particles into the atmosphere and noise are of less importance.

It is important to stress that the quantification of these aspects may vary from one facility to another depending on factors such as the size and age of the facility (Moletta R, Torrijos M, 1999), equipment, handling, cleaning plans, employee awareness, etc. ....

II.1.Water consumption :

Like most companies in the food industry, the dairy industry consumes large quantities of water on a daily basis (Table 2). This consumption is particularly high during auxiliary cleaning, disinfection and other operations.

Table 2: Quality aspects of water consumption in the dairy industry (UNEP, 2000).

Production process	Level of consumption	Most water-intensive operation	Comments
Milk	Bottom	Heat treatment Packaging	-
Cream and butter	Bottom	Pasteurisation of the cream Churning-Kneading	Washing the buttermilk before mixing
Yaought	Bottom	-	Especially in auxiliary operations
Ancillary operations	Student	-Cleaning and disinfection -Steam generation - Refrigeration	These operations involve the highest water consumption

II.2 Energy consumption :

The use of energy is fundamental to maintaining the quality of dairy products (Table 3), especially during heat treatment, cooling operations and product storage.

Table 3: Most common energy use in dairy enterprises (UNEP, 2000)

Energy	Most common uses	Equipment
Thermal	-Steam and hot water generation, cleaning	Pasteurisers/sterilisers, CIP cleaning systems
Electric	Cooling, lighting, ventilation, equipment operation	Electrically operated equipment (pumps, mixers, etc.), lights

This energy consumption during the various stages of production and packaging is also variable depending on the type of product being produced. It also depends on the characteristics of the factory.

II.3 wastewater :

The most important environmental problem of the dairy industry is the generation of wastewater, both because of its volume and because of the associated pollution load (mainly organic).The volume of wastewater generated by a dairy enterprise can range from 2 to 6 l/l of milk processed (UNEP, 2000).

cooling, condensate,

Table 4: Classification of wastewater generated in a dairy industry (Spreer, 1991).

Source	Description	Features	Volume

| Cleaning and process | Cleaning of surfaces, pipes, tanks, equipment. Loss of products, whey, brine, ferments, etc. | extreme pH, high organic matter (BOD and COD), oils and greases, suspended solids | 0,8-1,5 |
| Cooling | Cooling tower water, Condensate, etc. | T° variations, conductivity | 2-4 |

Table 5: Qualitative assessment of wastewater discharge in the dairy industry (UNEP; 2000)

Production process	Level of rejection	Operations generating the most wastewater	Comments
Milk	Medium	Heat treatment Packaging	Discharge is reduced by recycling the heat treatment water
Cream and butter	Medium	Pasteurisation Churning-Kneading Packaging	Buttermilk wash water has a high fat content
Yaought	Bottom	-	Mainly from auxiliary operations
Ancillary operations	Student	Cleaning and disinfection Cooling	The volumes and pollution load of cleaning water depend on the management of the water by the company. The discharge of cooling water depends on the degree of recycling of the cooling water.

Almost all of the COD (Chemical Oxygen Demand) in the wastewater of the dairy industry is attributable to the components of the milk itself, while impurities outside the milk itself are only a minority (about 10%).In fact, in addition to water, the composition of milk includes fats, proteins (in solution and in suspension), sugars, mineral salts, colouring agents and stabilisers, depending on the nature and type of product and the type of production technology used.

All these components are found in the wastewater in varying amounts after dissolving and being removed by the cleaning water.

In general, the liquid effluents of a dairy industry have the following characteristics:

- **High organic matter content**, due to the presence of milk components. The average COD of wastewater from a dairy industry is between 1000 and 6000 mg BOD/l.

- **Presence of oils and fats**, due to fat from milk and other dairy products (e.g. buttermilk wash water).

- High levels of **nitrogen and phosphorus**, mainly from cleaning and disinfection products.

- **Large variations in pH,** discharge of acidic and basic solutions. Mainly from cleaning operations.

- **High conductivity** (especially in cheese companies due to sodium chloride release from salting).

- Temperature variations (taking cooling water into consideration).

Milk losses, which can reach 0.5 to 2.5 % of the quantity of milk received or in the most unfavourable cases up to 3 - 4% (UNEP, 2000), constitute a considerable contribution to the pollutant load of the final effluent. 1 litre of whole milk is equivalent to a BOD5 of 110,000 mg O2/l and to a COD of 210,000 mg O2/l.

These losses can occur during the various production processes. There are many sources. They are summarised in Table 6

Table 6: Main sources of milk losses to wastewater (UNEP, 2000)

Process	Source of milk loss
Production of milk for direct consumption	-Leakage from storage tanks. -Overflow of tanks. -Leakage and leakage in pipes. -Deposits on the surface of equipment. -Removal of filter sludge/clarification. -Spills caused by damaged or deteriorating packaging. -Defects in the packaging line. -Cleaning operations.
Cream and butter production	-Flow during storage. -Leakage and leakage in pipes. -Overflow of tanks. -Cleaning operations.
Production of Yaought	-Leakage and leakage from storage tanks. -Outflows from incubation ponds. -Defects in the packaging line. -Cleaning operations

11.4 Waste:

Much of the waste generated in the dairy enterprise is inorganic in nature; it is mainly packaging waste (raw and secondary materials but also the final product). Other waste is also generated, related to maintenance and cleaning activities, or to the work in the offices and laboratory, which is either recycled or transported to a controlled landfill in order to reduce the pollution rate generated by this type of waste.

11.5 The impact of industrial discharges on the environment and the sanitation heritage:

> The impact on the environment:

The impacts of industrial discharges into the aquatic environment are mainly local and depend on the flow rate of the watercourse and the location of the discharge in relation to sensitive areas.

The food industry produces large amounts of organic matter which causes the proliferation of bacteria and other organisms, leading to a decrease in the oxygen content of the aquatic environment and the disappearance of species which have a high oxygen requirement.

The main effects of industrial discharges are localized on oxygen consumption (COD), eutrophication (nitrogen, phosphorus), and toxicity due to certain substances.

Industrial pollution emits many particles into the atmosphere and is responsible for 18% of greenhouse gas emissions. Heavy industry is responsible for the vast majority of these gases. They also produce a large proportion of acidic emissions, sulphur dioxide, nitrogen oxides, and carbon monoxide. They also produce emissions that contain volatile organic compounds and toxic substances (heavy metals, dioxins, various hydrocarbons...). Industries affect air quality and contribute to global warming.

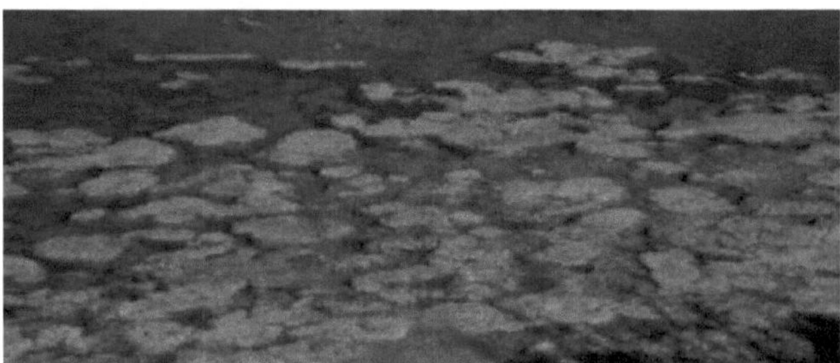

Figure 1: Eutrophication phenomenon

> The impact on sanitation assets:

Industries produce a wide variety of wastewater. Some are highly acidic and can destroy pipes if not neutralized before discharge. Others contain substances which can produce toxic gases dangerous to operating personnel or suspended matter which settles in the pipes and gradually clogs them.

At present, several wastewater treatment plants are the subject of complaints from residents and elected officials, in particular because of bad odours.

The acceptance of non-compliant discharges into the sewerage system can only aggravate these risks, particularly for suspended matter and sulphur compounds. Furthermore, the wastewater treatment plant makes it possible to reduce a domestic type of pollutant load or one that is similar to domestic waste. Contamination of wastewater with toxic substances

(hydrocarbons, cyanides, solvents, etc.) can alter the biological activity in the treatment tanks, greatly reduce the purification yield, and lead to
Non-compliant discharges from the wastewater treatment plant.

Figure 2: Discharge of industrial wastewater into the sewerage system.

III. The clean-up of industrial liquid waste :

Industrial discharges are much more variable in composition than domestic discharges. This variability is due to the sector of activity, the techniques used in the company, the production in progress and possible accidents or malfunctions encountered during production.

As the quantities discharged can be very large, their nature has a real impact on the sewerage system and the treatment plant. Therefore, the quality of the wastewater must be stabilised at a level acceptable to the public sewerage infrastructure before it is discharged into the sewerage system. This stabilisation of the quality of the discharged water at an acceptable level requires a minimum of works and operating rigour. Their use depends on the current production process. Each time the production process is modified, the wastewater discharges vary and the operation of the treatment works must be adapted to these new discharges. Inadequate adaptation of the operation of the wastewater treatment plant can lead to non-compliant discharges.

In the industrial sector, treatment methods depend on the type of industry and on factors related to the environment of the industrial establishment (e.g. available space). Thus, given the specificity of each industrial establishment, each wastewater treatment project must be the subject of a specific study to determine the most suitable wastewater treatment process or to review the industrial processes and the raw materials used in order to avoid pollution at the origin.

III .1 Basic principles of wastewater management in the companies :

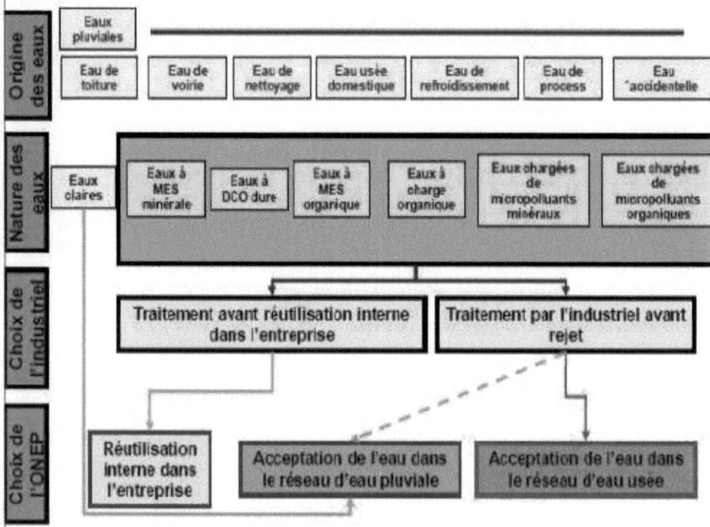

On examination of the scheme, it can be seen that there is a great diversity in the areas of production, nature and destination of the wastewater.

There are two possible approaches to their treatment:

- Collect all these waters and treat them together, before discharge into the sewerage system;

- Specific treatment of water that exceeds discharge limits at the point of generation.

In the current context, i.e. taxation of discharges based on drinking water consumption, the company's only objective is to discharge water of a suitable quality.

In this case, if the heavily loaded water is in the minority, the industrialist can use the dilution effect of the heavily loaded water by the lightly loaded water and thus achieve an acceptable water quality in the sewerage system. This approach requires a mixing tank before discharge. The volume of the basin will be determined on the basis of the daily discharge rate and the residence time required to ensure dilution of the loaded discharge before discharge.

In the context of load-proportional charging (which is in preparation) or the impossibility of diluting heavily loaded water with lightly loaded water, it is necessary to ensure partial depollution of discharges in order to comply with indirect discharge limits.

III .2 Industrial wastewater treatment techniques :

Generally, the treatment of small, heavily loaded flows is less expensive than the treatment of large, less loaded flows and filtration, decantation and flocculation techniques are more effective on fresh water than on water that has been in pipes and buffer tanks before being treated. Compact techniques for low cost pollution abatement are often :

- Degreasers (non-soluble greases);

- Desander and decanter for mineral particles above 200 urn;

- The sieves, including the rotating micro-sieves (from 0.1 to $10m^3$ /s) that can separate particles larger than 100 urn;

- Flocculation (to accelerate settling) with lime, aluminium or iron salts and polyelectrolytes (for particles between 1 and 1000 urn);

- Flotation, which separates particles using smaller structures than decanters; Membrane filtration, which concentrates the effluent and then treat ;

Evaporation, which is mainly adapted to very low flow rates;

Precipitation, particularly suitable for dissolved ions (cations and anions).

All the solutions that can be implemented either to ensure a discharge in conformity with the standards or to treat the water before its reuse are listed in the table below.

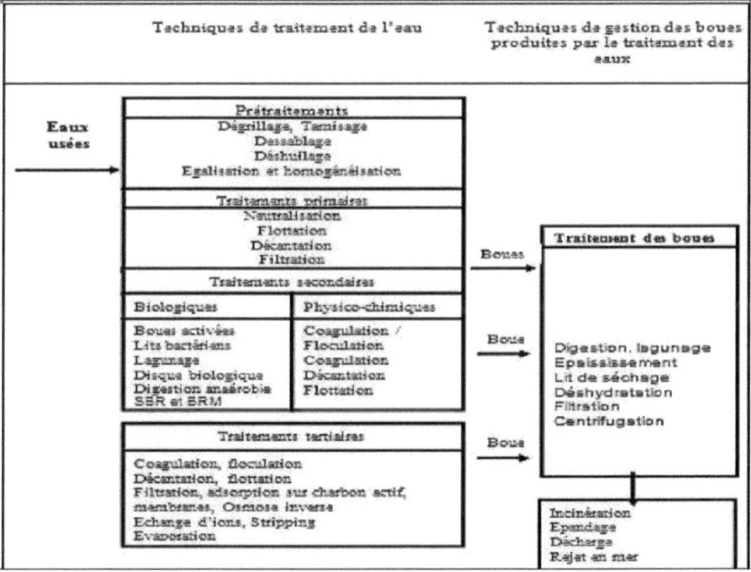

<u>Figure 3: Treatment solutions for industrial discharges</u>

> **Some techniques for biological treatment :**

Biological processes for pollution control are diverse. The treatment can be carried out by means of microorganisms fixed on a support or in suspension.

1. Biological treatment with fixed micro-organisms on a support:

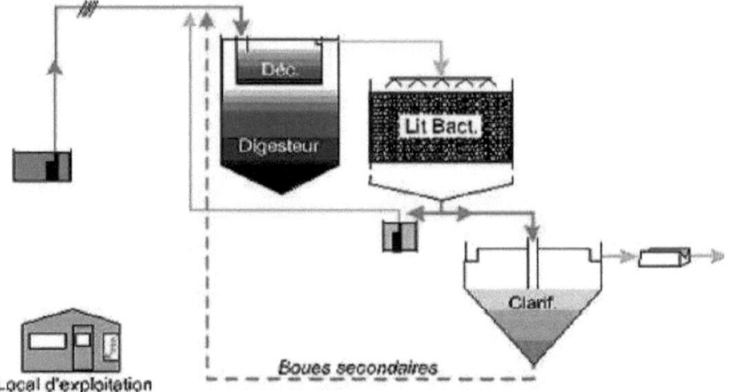

<u>Figure 4: Fixed crop treatment scheme</u>

1.1 Bacterial beds:

The bacterial bed is made up of a stack of inert materials of high porosity called "filter bed". The active mass of the purifying microorganisms is fixed on these supports whose rate of vacuum is close to 50%. The entry of water is always done at the upper part and the evacuation by the bottom because, in no case, the filter bed must be drowned (stop of the aerobic function) (Koudir a., Lamrani H., Louchli s, 1997).

The water is distributed as evenly as possible over the surface of the filter by sprinkling and aeration is generally done against the current by natural ventilation.

To meet the required discharge standards, the water must pass through the bed several times, which requires a recirculation pump. When the bacterial mass is too large, the bacterial film naturally breaks off and is separated from the effluent by settling.

❖ **Advantages and disadvantages:**

J Low energy consumption;

J Simple operation with little maintenance and control;

J Can be installed upstream of an active sludge plant to deconcentrate food industry type effluents;

J Good sludge settling.

❖ **On the other hand :**

J Low sensitivity to load variations compared to active sludge.

J Relatively high investment costs;

J Requires effective pretreatment;

J Sensitivity to clogging and cold ;

J Fermentable sludge.

1.2 Biofilters :

It is a purification process that combines biological degradation and filtration. The device consists of a decanter, a bio filter, a clean water tank and a dirty water tank. The oxygen requirements are met by air infiltration. Several times a day, a wash with treated water and air is necessary. Wastewater treatment consists of passing the wastewater through a tank containing a submerged filter material that supports bacterial growth. The biological filter therefore allows the assimilation of pollution and the filtration of the sludge produced, hence the term biofilter (Donkin M.J, 1997).

> **Advantages and disadvantages of the technique :**
> Good removal of SS;
> Low losses for the sludge produced,
> Compactness of the installation and better integration into the environment;
> Easy to automate.

❖ **On the other hand:**
> System not very suitable for concentrated effluents;
> High operating costs (energy) as well as investment costs;
> Sensitivity to sudden changes in hydraulic load and clogging;
> High consumption of treated water during washing.

2. Biological treatment with suspended microorganisms:
2.1 Lagooning :

It is a process based on a natural biological equilibrium involving; (i) bacteria which degrade organic matter, (ii) algae which provide oxygen by photosynthesis, (iii) zooplankton which consume the algae and sometimes reeds which filter the water. The system consists of a series of three lagoons arranged in a row, sized in total for a residence time of the wastewater of at least 150 days.

- The first allows the abatement of carbon pollution.
- The second allows for the removal of nitrogen and phosphorus.
- The third refines the treatment and makes the system more reliable in the event of a malfunction in an upstream basin or during a maintenance operation.
- **Advantages and disadvantages of the technique :**
- Limited investment costs (in the absence of strong sealing constraints).
- Low operating costs.
- Good integration into the environment.
- Good elimination of pathogens.
- Sludge with low fermentability.

- Good nitrogen (70%) and phosphorus (60%) removal.

❖ **On the other hand:**

- Large footprint;
- Soil and waterproofing constraints;
- Seasonal variation in treated water quality ;
- Nuisances if defective in design and/or operation (rodents, odours, etc.);
- Sludge extraction difficulties ;
- Sensitivity to septic effluents and concentrates;
- The quality of the waste water is lower than that of conventional processes.

Figure 5: Natural lagoon treatment plant.

2.2 The Sequential Biological Reactor (SBR):

The SBR technique consists of purification of the waste by active sludge in a sequentially fed and aerated reactor (Petillot F, 1974).

The system consists of a wastewater reception tank, an aerobic treatment tank and fine bubble aerators fed by a blower. Two submerged pumps allow the evacuation of the treated effluent at the end of the decantation phase and the removal of excess sludge (Garrido J.M, et al, 2001).

This method is characterised by its simple operation and excellent purification performance.

2.3 Active sludge :

❖ The operation of the activated sludge process:

The treatment process is called "active sludge" because all the conditions favourable to maximum bacterial activity are implemented: a sufficient supply of oxygen, a supply of nutrients if the effluent does not contain all the compounds necessary for the development of the bacteria, permanent agitation in order to promote contact between the bacteria and the pollution, and a high concentration of bacteria in order to increase the efficiency of the treatment. The treatment chain is composed of a bioreactor, a clarifier and a sludge recycling loop.

❖ **Advantage:**

A proven process for achieving the highest processing performance.

Suitable for high organic loads.

Process suitable for the treatment of phosphorus.

Procedure adapted for separate or combined sewers associated with a storm water basin.

Limited land   area.

❖ **Inconvenient:**
Rigorous   operation (electromechanical monitoring).
   High operating costs   especially for small installations,
4 to 8 % of the investment cost annually.
 Training of staff for operation.
 Consequent sludge production requiring an adapted treatment according to the capacity of
the works.
 Higher energy costs than for a rustic system.

III.3 Industrial clean-up: Open financial opportunities.

The development of the national industrial sector has given a boost to the national economy, but in some areas it has contributed to environmental degradation, thus compromising the sustainable development to which our country aspires. In order to counter this trend, Morocco has set up, with the support of the German cooperation, the Industrial Depollution Fund (FODEP). FODEP is an incentive instrument that encourages environmental upgrading through technical and financial support to industrial enterprises:

■ The installation of facilities for the reduction and elimination of all forms of liquid, solid or gaseous pollution.

■ The realisation of projects ensuring the economy of resources, especially water and energy, through the change of processes and the use of clean technologies.

Since its creation, the FODEP has agreed on about a hundred depollution projects for a total amount of nearly 500 MDH, of which 190 MDH represent the State's counterpart, which is financed by donations from the KFW (Credit Institution for Reconstruction)

■ **Mechanisms for industrial water pollution control :**

■ Maintenance and improvement of the budgetary effort made by the State to finance the voluntary mechanism inspired by the experience and results of the FODEP to meet the needs of industrial pollution control (ID)

■ The River Basin Agencies will be responsible, in consultation with local stakeholders, for

■ identification of the needs in terms of industrial water pollution control ;

■ assistance to promoters selected to benefit from HID incentives;

■ monitoring the execution and implementation of industrial clean-up projects

water ;

■ the verification of compliance with the agreed objectives and;

■ the payment of grants to beneficiary companies in accordance with well-defined criteria.

■ **Eligibility criteria :**
Companies meeting the following criteria are eligible for HBS support:
- Registration in the Commercial Register
- Regularity of the situation with regard to the tax authorities and the CNSS.

- All non-domestic projects are eligible.

- No restrictions on the cost of the project.

- The subsidy ceilings are fixed at a maximum of 5 MDH for individual projects and 10 MDH maximum for collective projects.

- The maximum subsidy per project type is 40% for downstream projects and 20% for integrated projects.

- To have the declaration and authorisation for discharge from the ABH, in accordance with articles 52 and 53 and the provisions of law 10/95 on water.

Specific agreements must be drawn up between the Department of the Environment, the ABH concerned and the beneficiary company, setting out the nature of the ABH's intervention.

IV. Regulatory context :

In terms of regulation, it is essential to distinguish between discharges into a public sewerage system and discharges into a natural receiving environment.

In the case of a discharge into a natural environment, the industrialist is subject to the law 10-95 on water and its application decrees published or in the course of preparation.

In the case of a discharge into a public sewerage system, the industrialist, as a user of the public sewerage service, is subject to the local regulations concerning the use of the public sewerage service. A series of clauses in the contract signed with the municipality define the terms of use and operation of the public service. These clauses have regulatory value and are enforceable against and by third parties. This means that the industrialist has to comply with these prescriptions but also that any user can complain about the non-respect.

In addition to these regulatory tools, the legislation on classified establishments and the deeds of sale of plots of land intended for industrial activities also contain provisions relating to wastewater discharges and environmental protection, but do not allow for direct intervention by the public sewerage service operator, unlike the contract between the municipality and the sewerage system operator.

Therefore :

- companies that discharge directly into the receiving environment must comply with the law 10-95 on water and its application decrees and the sewerage system manager does not intervene in the management and monitoring of these discharges;

- companies discharging wastewater into the sewerage system managed by the sewerage system operator must comply with the requirements of this service, the main points of which are as follows:

- only install and use connection boxes approved by the sewerage system operator;

- declare connections and discharges to the sewerage system operator ;

-   comply with the following discharge limit values (DLV):

-   VLR "direct discharge" for discharges to the stormwater system;

-   VLR "indirect discharges" for discharges into the wastewater system.

The tables with the discharge limits are shown below:

Table 7: Discharge limit values (Ministry of the Environment of Morocco, 2002)

	VLR direct	Indirect VLR
Temperature : °C	30	35
PH	6,5-8,51	6,5-8,51
BOD5 mg of 0 /ml$_2$	100	500
COD mg of 0 /ml$_2$	500	1000
TSS mg/l	50	600
Conductivity in ps/cm	2700	
Ag mg/l	0,1	0,1
Al mg/l	10	
As mg/l	0,1	0,1
Ba mg/l	1	1
Cd mg/l	0,2	0,2
Co mg/l	0,5	1
Cr mg/l	2	2
Cr VI mg/l	0,2	0,2
Cu mg/l	0,5	1
Fe mg/l	3	3
Hg mg/l	0,05	0,05
Mn mg/l	1	1
Ni mg/l	0,5	0,5
Pb mg/l	0,5	0,5
Sb (Antimony) mg/l	0,3	0,3
Se mg/l	0,1	1
Sn (tin)	2	2
Zn mg/l	5	5
Free cyanides (CN) mg/l	0,1	1
Free sulphides (S2-) mg/l	1	1
Fluorides (F) mg/l	15	15
Sulphates mg/l		400
Hydrocarbons mg/l	10	20
AOX	5	5
Phenol index mg/l	0,3	5
Detergents	3	
Oils & Fats	30	50
Kjeldahl nitrogen mgN/l	30	
Total phosphorus P mgP/l	10	10

Derogations to these discharge limits may be considered provided that the quality remains compatible with the sanitation processes implemented.

This type of derogation can be considered on the basis of a characterisation of the discharge and the operating data of the public sanitation service.

Part II
Presentation of the study area

I. General presentation of the Cooperative Laitiere du Maroc Oriental (COLAIMO):

I.1 Portrait of the cooperative:

The main information about the cooperative is summarised in the table below:

Information	Dairy
Date of creation	1956
Investment	242,000,000 DH
Capacity :	150,000 litres/day
Range of products manufactured	Pasteurised milk, fermented milk, yoghurt, lben and butter
Collection areas	Oujda, Berkane, Nador, Taourirt, Guercif and Jerada.
Number of employees	300 people
Walk	Market covering mainly the region

I.2 The location of the cooperative:

The factory is located in the industrial zone bordering the city of Oujda. The area where the dairy is located was initially rural. Today, it is home to an intense urban activity. A large urban development project is underway. The satellite photo below gives an idea of the location of COLAIMO.

Figure 6: Aerial view of COLAIMO

1.3. Manufacturing processes :

The recently rehabilitated dairy is equipped with modern machinery and equipment. It is supplied by eight tankers belonging to the cooperative itself.

a. Industrial production :

COLAIMO consists of 4 blocks: pasteurised milk, Leben, butter cream and desserts. On average, the dairy recycles 120 tons of milk per day and produces about 80 tons of pasteurised milk and the rest is distributed to the other derived products.

> **Apergu technology for pasteurised milk :**

After being normalised to the required fat content, the milk is passed through heat exchangers for pasteurisation. COLAIMO has two pasteurisation lines and processes an average of 80,000 l/d of milk. The packaging machines ascetically package the pasteurised milk in half-litre flexible plastic bags. Here are the different operations that the pasteurised milk undergoes:

> **Cooling:** The temperature of the milk is reduced to a tolerable level. The temperature of the milk must be less than or equal to 6°C. The milk must be kept at the same temperature during storage.

> **Mechanical clarification:** Before processing, the milk must be mechanically clarified. This is done at the reception area. A centrifugal separator is used to perform this operation

> **Thermisation:** Thermisation is carried out in a plate heat exchanger. The milk is heated to a temperature between 57 and 68 °C with a maximum holding time of 30 s. Thermisation is often used to improve the storability of milk before processing.

-Fat standardisation: The fat content of milk is subject to significant fluctuations. Skimming is a mechanical separation of the cream which allows the fat content to be brought down to the regulatory level of 3%. The operation is carried out in a skimming machine. The optimum temperature for skimming is between 50 and 60 °C.

-Homogenisation: Homogenisation **is** mainly used in the dairy industry to prevent the separation of fat globules by reducing their diameter to 0.5 - 1pm which increases their number and surface area by a thousand times. The optimum operating conditions are between 60 and 70°C for temperature and 130 and 140 bar for pressure.

-Heat treatment: The main purpose of heat treatments is to kill pathogenic microorganisms in the milk. For example, with a pasteurisation temperature of at least 72 °C for 15 seconds or another combination of temperature and treatment time that results in efficient pasteurisation (negative phosphatase test and positive peroxidase test).

> **<u>Yoghurt and Dessert Technology Apergu</u>**

The preparation of these products uses milk, milk powder, water, sugar and flavourings. The resulting mixture is pasteurised before being seeded. The fermented product may undergo further processing before being packaged in plastic jars.

> **<u>Milk powder :</u>**

The excess milk collected during the high lactation period is sent to Kenitra to be processed into milk powder.

b. Marketing :

The distribution radius has recently undergone an important expansion which has allowed the cooperative to increase its share of the national market. In addition to the market in the Oriental region, the distribution network extends to the regions of Taza Al-Hoceima, Fes, Meknes, Er-Rachidia and Rabat.

11.   Description of the sectors :

II.1 Water supply :

The water consumed by COLAIMO comes from the RADEEO network and from a deep borehole (28 to 30 m) and undergoes a complete treatment in the cooperative's water treatment plant.

The borehole water is treated as follows:

- Sand filter (2 tanks operating in "bachee" mode with backwash).

- Osmosis machine based on the principle of reverse osmosis.

- Coagulant injection (alumina silicate upstream of filtration)

- Buffer tank of about 200 m3

- Pumping and denitrification on ion exchange resins (anionic +cationic) input 36 mg/l of N

NO3- output at 26 mg/l

- Softening to 0° Frangais of part of the water and mixing to obtain 23° Fr.

On the whole flow (Filtration on polypropylene membranes).
-Organic disinfection by addition of peracetic acid
- Storage in a 500 m3 tank
- Pumping and pressure boosting
- 50 micron microfiltration
- Chlorination by bleach injection up to 0.7/0.8 PPM
- Dechlorination on activated carbon
- 4 bar overpressure

- Microfiltration at 3 microns before delivery to the production workshops **All the buffer volumes upstream of the production process represent 700 m3 E.g.: Approximately one day's production**

 Sanitary water :

Use of city water.

Forage

Osmoseur

Filtres à sables

Adoucisseur

Drilling   Osmosis

Sand filters Water softener

11.2   Water distribution scheme :

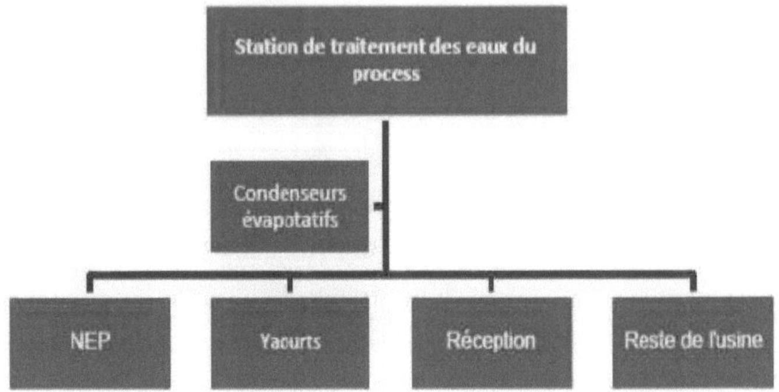

The treated water is sent to a large 500 tonne capacity tank for storage. The various workshops are supplied with water from this tank via a main pipeline equipped with a general meter. The daily readings from this meter are recorded in a register. There are also two other meters installed on the networks supplying the cold production workshop and the second CIP used in the cleaning of the downstream part of the production line. It should be noted that the factory has a second CIP used for cleaning the downstream part of the production line.

11.3   General water consumption of the plant :

The daily water consumption of COLAIMO is about 310 m3/d. If we compare this consumption with the average daily volume milked over the same period, the specific consumption of COLAIMO is about 4 litres of water/litre of milk milked.

The water consumption workshops are :

Cleaning of equipment and premises

Steam, hot and cold water production

The making of yoghurt.

Cooling of machines

Condensation of refrigerants

Watering of green areas and,

Other

Table 8: Summary of water consumption at the plant (COLAIMO, 2009).

Reason roe	Reception	Manufacturing	Packaging	C.I.P	Ctiaudiere+Cold	Sanitary
Treated borehole water	out	yes	yes	yes	Yes + specific adouclssemefit	City water
Debrt/day	60 rn3/d	100 m3/d	10m3/d	400m3/d	30 m3/d	negligent
DetMtfli estimation	B m3/h	9 m3fti	1 to 2 m3fti	140 гиЗЛт	30 m3fti	

☐ebitfj medium	**600** ni3/Day
Debrt/d peak y including 20% for the future	1000 ип3/day

II.4 Quantification of discharges :

1. Product (milk) losses :

Approximately 2% of the milk received is lost in the various operations that take place inside the plant, i.e. about 2.4 tonnes/day. Two kinds of losses are to be accounted for: (1) losses due to the material that remains stuck to the walls of the equipment after each operation and that goes down the drain during cleaning operations, and (2) losses generated by the pushing operations carried out in the vats and tanks at the end of each emptying. Annually, COLAIMO loses a little less than 900 tons of milk, which constitutes an annual loss of income of a little less than five million dirhams.

2. Liquid effluent from the dairy:

The quantification and characterisation of discharges is essential to assess the water savings to be achieved and to possibly size the resulting projects.

The discharges identified are:

 Equipment cleaning water;

 Floor and tanker cleaning water ;

 Machine cooling water;

 Condensation water from heating vapours ;

 Cooling water used in refrigeration circuits ;

 Shoot water or white water;

It should be noted that the cleaning of equipment is the major source of liquid discharges. After each treatment operation, the equipment is cleaned as follows:

❖ Washing with water to remove traces of milk remaining inside the equipment;

❖ Washing with dilute soda solution;

❖ Ringage a l'eau ;

❖ Nitric acid wash; and

❖ Ringage a l'eau.

> Water from Cleaning in Place (CIP):

The CIP units used to clean a large part of the dairy's workshops consist of 4 tanks connected in a closed circuit to equipment which is cleaned after each operation. The soda and acid solutions prepared and stored in two different tanks in the CIP unit are used for several cleanings and are only sent to the sewer when they are exhausted. It should be noted that the water from the last ringings is only partially recovered and used in the first washing during the cleaning that follows later. Some equipment is still cleaned in open circuit. While the cleaning operations are the same as for CIP, no recycling of acid and soda solutions or ringing water is carried out.

The CIPs are responsible for cleaning the 35 circuits in the dairy. Based on the flow rate of the CIP pump (20 m3/h) and the duration of the pre-washing and ringing phases, more than 140 m3/d of water is discharged to the sewer. More than half of this volume is clean as it is ringing water.

The cleaning of the pasteurisers operating once a day results in an evaluated discharge based, as before, on the flow rate of the pumps and the duration of each phase. The result of this evaluation is recorded in the table below:

<u>Table 9: Cleaning of the two pasteurisers.</u>

Pasteu risateur	Pump flow	Phase	Discharge per phase m3 /D	Total discharge m3 .'J
Paste* milk		Prewash Ring^ge 1 Ringage 2	1.5 1.5 1.5	4.5
Paste* of derivative products		Prewash Ring^ge 1 Ringage 2 Sanitation	1 1 1 1.5	4.5

The cleaning of the two pasteurisers generates water discharges of around 9 m3/d.

- **Cooling**

Only the packaging machines and the skimmer are continuously cooled. It was noted on site that the skimmer is cooled in a closed circuit and consequently does not cause any significant loss of water. As for the filling machines, which operate 12 hours a day and are cooled by means of a water flow rate of about 1.5 litres per minute, their water consumption is estimated at about 1 m3/d.

- **White water**

Also called push water because after emptying the tanks or cisterns, the water is used to push the milk through the circuits. Some of the milk mixed with the water is sent directly to the sewer. This is one of the main sources of milk loss during the manufacturing process. This water is rich in biodegradable material and can be used as a reusable raw material in other sectors such as animal feed production.

- **Cleaning the floor**

Cleaning is carried out after each operation. It is noted that the use of the self-cleaning trolley minimises water consumption during cleaning.

- **Washing of boxes**

The crate washer consists of three tanks. The first two tanks are heated with steam. The temperatures are 40°C and 70°C respectively. Although the consumption of this machine is not significant, improvements in operation and design can significantly reduce it. The repair of leaks and the recycling of water from the last tank can save about ten cubic metres per day.

CIP production

Milk dumping and truck washing station

■ **Summary of discharges :**

The quantities of the various effluents discharged as assessed over a 24 hour period are recorded in the table below:

Table 10: Average daily discharge from the dairy (COLAIMO, 2009).

Nature of the Discharge	Source	Quantity (m3/d}
Clean IP discharges	Cooling water for cooling machines and other equipment Condenser feed water and cooling tower blowdown water Condensate	5 20 10
Hairy rejects	CIP system cleaning water Cleaning water CIP pasteurisers City cleaning waters Growing waters Lower butter factory product Cleaning the floor Washing of boxes Other	140 9 12 20 5 2 5 10
Totawc Clean Discharge		35
Total discharges Polities or supposedly polluted		202
Total assessed releases		237
Average total consumption based on discharges plus 30 S_.		310

3. Pollution load of discharges :

The pollutant load of the dairy effluent is a crucial parameter in the choice of the effluent reuse option. Dairy effluent usually contains sugars, proteins, fat and residues of additives (colourings, etc.). The environmental guidelines for liquid dairy effluent from the good practice manual on pollution prevention and control (World Bank) are as follows

Table 11: Minimum pollution loads

Parameters	VaJeitrs min [mates
BOD5	50 nig-'l
DC-O	250 mfj.'l
MY	50 nigTl
PrtZ-tc	2 Tig/l
MT K	IO
Mui les and Crraisses	IO nig-'1

These values are closely related to the amount of milk lost in the system. Knowing that one litre of lost milk can generate 110 kg of BOD in the effluent, a theoretical calculation shows that in COLAIMO's effluent, one can find at least 500 mg/l of BOD, which is about ten times the recommended value.

4. Summary of dairy discharges :

❖ Daily water consumption of the dairy: 310 m3/d

❖ Specific consumption: about 4 litres of water / litre of milk

❖ Daily volume of liquid discharges from the dairy: approximately 240 m3/d

In the following, the techniques used to characterise the pollution at different levels of the milk production process and its by-products are described. The physico-chemical and microbiological analyses of the liquid effluents studied were carried out in the laboratory of the Oriental Centre for Water Science and Technology and in the Regional Analysis and Research Laboratory of ONSSA.

The determination of the physico-chemical and bacteriological composition of a wastewater is essential to assess its quality and to characterise it as well as possible. As it is necessary to operate on the sample, it is important to limit as much as possible the errors associated with each of the following steps:

- Taking the sample ;
- The conditions and duration of its storage before analysis ;
- The analysis itself, in the laboratory or in the field.

The AFNOR standard expresses well the essential principles for this type of analysis: "A sample cannot be taken just anywhere, just anyhow and by anyone. A sample can only be taken according to the objective pursued. The choice of locations where samples are taken, the number of samples taken, the manner in which samples are taken, processed, preserved and transported depend on the information sought and essentially on the analyses to be carried out.

1. <u>Collection of samples :</u>

The sampling is an important operation, it conditions the analytical results and their interpretation. The physico-chemical and bacteriological characteristics of the water must not be altered.

The wastewater sample taken should allow for analysis to better understand :

- the different types of pollutants released,
- their respective concentrations,
- possible technical solutions to reduce or eliminate pollutants.

A sample must be REPRESENTATIVE of the total discharge from the plant or a sector of the plant.

The analytical samples were taken from the main collector (CP) receiving the overall raw dairy effluent located downstream of the dairy unit under study, during a normal period of operation of all the workshops of the audited unit and which covers in terms of sampling time and frequency all the activities and types of raw material processing. The purpose of these samples is to monitor the evolution of the physico-chemical and bacteriological parameters of the dairy effluents in order to characterise the pollution of the latter.

2. <u>Storage and transport of samples:</u>

In general, as little time as possible should elapse between the collection of the sample and its analysis. Depending on the nature of the assay, special precautions in the handling of the sample are necessary, in order to avoid the rapid evolution of the water over time which may induce natural interferences such as microbial growth and the loss or gain of dissolved gases...etc. To this end, the water used for the analysis of physico-chemical and microbiological parameters was transported at 4°C in portable ice boxes.

3. The main parameters and analytical methods:

The maintenance of a good hygiene of the dairy is dependent on numerous cleaning operations of the premises as well as of the equipment (tanks, pipes,...) and this, as well with cold as hot water in addition sometimes of various additives (chemical products).The effluent water contains various constituents quantifiable by the means of the measurement of various parameters which make it possible to characterize the pollution. The parameters measured are temperature, pH, conductivity, dissolved oxygen, suspended solids (SS), biochemical oxygen demand (BOD5), chemical oxygen demand (COD), total phosphorus (TP), ammonium (NH_4 +) and orthophosphates.

- **The temperature :**

The temperature of the milk effluent was measured on site at the time of sample collection with an accurate thermometer, calibrated to 1/10th of a degree.

- **The pH:**

This parameter was determined electrically using a pH meter type WTW 197i. The accuracy of the measurement given by the manufacturer is ± 0.1 pH unit. The use of this characteristic is important because it has a strong influence on the possible use of microorganisms in pollution control systems. In addition, discharges that are too alkaline or too acidic could have a negative impact on the environment.

- **Electrical conductivity :**

It is measured using a WTW conductivity meter, model 330i/SET, and the values are obtained with a margin of error of 2%.

- **Dissolved oxygen :**

The amount of dissolved oxygen is measured in mg/l by the Winkler method, which is based on the oxidation of manganese hydroxide in a strongly alkaline solution. By acidification in the presence of an iodide, the manganese hydroxide formed is then dissolved. This results in the release of free iodine in an amount equivalent to that of the oxygen initially dissolved in the sample. This free iodine is titrated by means of a standard solution of sodium thiosulphate N/80 using starch as an internal indicator after most of the iodine has been reduced (AFNOR T90).

- **Suspended matter :**

Suspended solids are all the mineral and organic particles contained in wastewater. Their effects on the physicochemical characteristics of the water are very harmful (modification of the turbidity of the water, reduction of the penetration of light endangering photosynthesis).

The experimental protocol for the measurement of TSS is very simple; the samples were filtered on a 0.22 or 0.45 micrometre porosity membrane, dried at 105°C and weighed to determine the level of suspended solids in the dairy effluent.

- **<u>Biochemical oxygen demand (BOD^):</u>**

The measurement of the Biochemical Oxygen Demand after 5 days was facilitated by the use of Oxitops WTW (according to AFNOR NFT 90-103).

<u>Figure 7: BOD metre Oxitops WTW</u>

The decrease in oxygen consumed during the biodegradation of a sample causes a decrease in pressure measured with a manometer. The water sample is introduced into a thermostated chamber and incubated in the presence of air.

The micro-organisms present consume dissolved oxygen which is continuously replaced by oxygen from the air volume above the sample. The carbon dioxide formed is trapped by sodium hydroxide. In a vial containing a magnetic rod, the sample, containing if necessary a bacterial plating, is introduced. The volume introduced into the vial depends on the DB05 measured and the apparatus used. In the scoop to be placed on the vial, 2 to 3 grains of sodium hydroxide are introduced to which 1 to 2 drops of water are added (take every precaution to ensure that the solution does not overflow and contaminate the sample). The vial is then tightly closed. Then incubate at 20°C for 5 days with constant agitation. Follow the manufacturer's instructions for data acquisition.

- **<u>Chemical oxygen demand (COD):</u>**

The Chemical Oxygen Demand (COD) has been calculated according to the AFNOR standard (NFT 90-101) and expresses the amount of oxygen required to oxidise the organic matter (biodegradable or not) of a sample with an oxidant, potassium dichromate ($K_2 Cr_2 O_7$).

Under defined conditions, certain materials contained in water are oxidised at boiling point (150°C) by an excess of potassium dichromate, in an acid medium and in the presence of silver sulphate acting as an oxidation catalyst and mercury (II) sulphate allowing the chloride ions to be complexed. The excess of potassium dichromate is dosed by iron ammonium sulphate (Mohr's salt) in the presence of feroine as an indicator. The colour changes from green to brick red. The oxidizable materials (and in particular the organic materials) in the sample are oxidized by potassium dichromate under the specified conditions.

Potassium dichromate is reduced:

$$Cr_2O_7^{2-} + 14H^+ + 6e^- \rightarrow 2Cr^{3+} + 7H_2O$$

The residual potassium dichromate is dosed with a solution of iron (II) ammonium sulphate (i.e. Fe 2+), in the presence of feroine (redox indicator):

$$Fe^{2+} \rightarrow Fe^{3+} + e^-$$

The overall reaction of the assay is as follows:

$$Cr_2O_7^{2-} + 14H^+ + 6Fe^{2+} \rightarrow 2Cr^{3+} + 6Fe^{3+} + 4H_2O$$

It is then possible to determine the amount of potassium dichromate consumed during the test and to deduce the equivalent amount of oxygen. To limit the interference of chlorides, mercury sulphate is added, which leads to the formation of chloromercurate (II), which is soluble and not very oxidisable:

$$Hg^{2+} + 2Cl^- \rightarrow HgCl_2$$

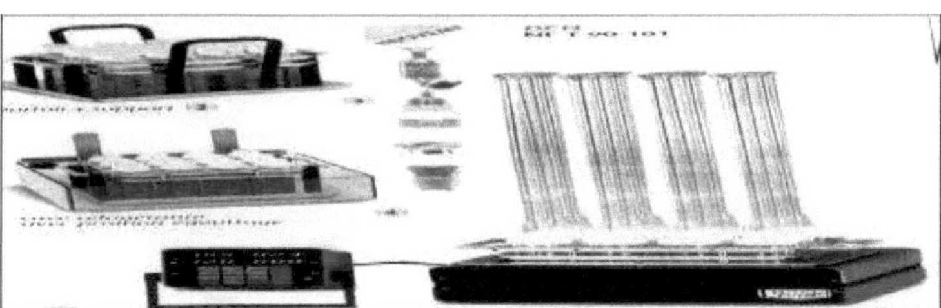

Figure 8: COD Mineraliser

■ **Total phosphorus (TP):**

The determination of total phosphorus is measured according to the standard (AFNOR T 90). After hot mineralisation of a test sample in the presence of sulphuric acid and perchloric acid, the orthophosphates obtained are then read by spectrophotometer at a wavelength of 880 nm.

■ **Orthophosphates :**

The orthophosphates were measured according to the standard (AFNOR T 90).in an acidic medium, a complex of ammonium molybdate and potassium antimony double tartrate is formed. This complex is then reduced by ascorbic acid resulting in a blue coloration which has a maximum absorption value at a wavelength of 880 nm in spectrophotometry.

■ **Ammonium (Ntf$_4$ +)**

The determination of ammonium is carried out according to the AFNOR standard (NFT 90-

105). To 20 ml of water to be analysed are added 1 ml of sodium nitroprusside and phenol solution and 1 ml of chlorine solution. The vials are then shaken and placed in the dark for at least 6 hours. The reading is taken on the spectrometer at a wavelength of 630 nm, taking into account the value read for the control (20 ml of distilled water instead of the water to be analysed). Refer to the calibration curve to obtain the concentration in mg.l-1.

4. **Microbiological characterisation :**

The microbiological analyses consisted of a search and enumeration of the main indicators of pollution and possible faecal contamination.

■   **FMAT count (Total Mesophilic Aerobic Flora):**

It is an aerobic flora enumerated by petri dish culture on selective gelose medium for enumeration. 1 ml of the sample to be treated (stock solution) is placed in the petri dish, the inoculation is done with a single layer of gelose, then incubated in an oven at 30°C for 72 hours. The enumeration is done by counting the colonies (expressed in CFU/ml).

■   **Yeast and mould counts :**

Yeasts form round, white, opaque colonies, while moulds form round, filamentous colonies, varying in colour from white to green. The medium used for the culture of yeasts and moulds is potato extract glucose gelose. 1 ml of the stock solution (sample) is introduced into the petri dish, the inoculation is done by a single layer of gelose. The reading is taken after incubation for 5 days at a temperature of 30°C.

■   **Coliform counts :**

Coliforms are non-sporogenic, Gram-negative, oxidase-negative, facultatively anaerobic organisms, and have the same specific properties after incubation at 44°C, hence the name **"thermotolerant coliforms"**.

The medium used for coliforms is tergitol with TTC (triphenyltetrazolium chloride). Petri dishes are incubated in an inverted position at 37°C for total coliforms and at 44°C for faecal coliforms for 24 hours. Confirmation is done with the bright green lactose broth. The principle of this medium is based on the ability of coliforms to ferment lactose with gas production.

■   **Staphylococcus pathogen enumeration :**

Coagulase-positive staphylococci are characterised by the formation of grey or black colonies surrounded by a clear, stable, opaque halo of fibrin.

The culture medium used for staphylococci is BP (BAIRD PARKER). The staphylococci petri dishes are read after 48 hours of incubation at 37°C.

Bacteria secreting a heat-stable toxin. The pathogenic risk from coagulase-positive staphylococci arises if their presence exceeds the norm. This level is necessary for the toxin responsible for the poisoning to be in sufficient quantity to trigger either sudden disorders such as vomiting, nausea, abdominal cramps, or severe disorders in children, the sick and the elderly.

■ **<u>Salmonella count</u>** :

They belong to the Enterobacteriaceae family and are motile, Gram (-), aerobic and optionally anaerobic rods. They ferment glucose, maltose and mannitol, producing gas, but do not ferment sucrose. They reduce sulphite to sulphide and decarboxylate lysine. The culture media used are green bawling gelose (GVB) and Xylose-Lysine-Desoxycholate (XLD), and the counts were made after incubation of RVS medium at 42°C and MKTTn medium (MULLER-KAUFFMANN broth with Tetrathionate-Novobiocin) at 37°C for 18h.

Part IV
Analysis results and discussion

I. **The origin of wastewater :**

Seven main sources of liquid discharges have been identified within the plant during the operations of :

• **Reception site**: After each cleaning phase, DIVOSAN is poured into the milk storage tanks. This product is discharged into the sewage system directly before the tanks are filled.

• **Pasteurisation:** When a milk storage tank at the receiving site is emptied, the pasteurisation cycle can be interrupted while waiting for the milk to arrive, in order to avoid heat damage to the pasteurisation plant, clear water is sent through the circuits specific to incoming milk and then discharged directly into the sewer.

• **Butter production**: The churning of butter and its solidification produce large quantities of water that are discharged directly into the sewer system without any treatment.

• **Boiler blowdowns**: Steam for a variety of uses is obtained from the boiler. The boiler generates effluents with varying loads depending on boiler maintenance, contact of the steam with surfaces containing chemicals, anticorrosive and antioxidant agents and additives added to the boiler water.

• **Cleaning of the premises:** this is water from the cleaning of the premises and machines during and after each production run, and also product losses from filler adjustment and the extent of cylinder explosions, especially during the production of gaseous products.

• **Cleaning of the water treatment line filters***:* the filters are backwashed with clean water.

• **CIP (Cleaning in Place)***:* The CIP system is a set of self-contained components that allow the right cleaning and sanitising solution to be circulated through non-demolished equipment in the right concentration, at the right place, at the right temperature, at the right flow rate, with the right mechanical action and for the right contact time. The right combination of these factors will ensure the effective cleaning and sanitising of food contact surfaces. The cleaning cycle consists of the following steps:

• **Preliminary ringing**: a preliminary ringing is necessary to remove the
non-adherent stains. For this purpose, clear water (from the well) is used or recovery water from future ringings.

• **Alkaline cleaning**: a soda ash cleaning is performed to remove organic soils, the solution is then recovered for reuse.

• **Inter-ringing**: this first ringing is designed to remove alkaline detergent residues, the water collected in a buffer tank is channelled via a suitable pipe to the sewer.

• **Acid cleaning**: A wash with a hydrochloric acid solution is used to remove mineral soils and to remove and neutralise any traces of soda. The solution is then recovered for reuse.

•   **Inter-ringing**: This second ringing with clear water is designed to remove traces of the acid detergent. The water from the end of this rinse is recovered and used for the preliminary rinse of the next equipment to be washed.

•   **Sanitation**: sanitation is carried out shortly before production operations, the chemical used is active DIVOSAN, the purging is done before the product is pumped.

II. Values of physico-chemical and microbiological parameters of the cooperative's dairy effluent at the main collector:

II.1 Results of chemical analysis :

The results obtained (Table 12) show that several physical parameters (TSS, T°C,) as well as chemical parameters (BOD$_5$, COD) are high and exceed by far the values authorised by the Moroccan standard on direct discharges.

Table 12: Values of physico-chemical parameters at the collector main liquid discharges from COLAIMO

Parameters	Value found	Normative value
T°C	35°C	30
pH	8,5	6,5-8,51
Conductivity (gs/cm)	3027	2700
TSS (mg/l)	300	50
O_2 dissolved (mg/l)	2	Not identified
BOD$_S$ (mg/L)	2300	100
COD (mg/l)	3900	500
PT (mg/l)	14	10
PO$_4$ 3- (mg/L)	9,5	Not identified
NH$_4$ +(mg/l)	29.5	Not identified

11.1.1.   The temperature :

The value of the temperature measured at the level of the main collector of the cooperative exceeds the temperature authorised by the Moroccan standard on direct discharges (30°C). This is mainly due to the cleaning water of the CIP station, this high temperature is due on the one hand to the exergonic reactions linked to the addition of acid and soda during these cleaning operations, and on the other hand to the use of superheated water.

Water temperature is an ecological factor with important ecological implications. It affects the density, viscosity and solubility of gases in water, as well as chemical and biochemical reactions, the dissolution of dissolved salts, and the development and growth of living organisms in water, especially micro-organisms (WHO, 1987).

Temperature influences the amount of oxygen dissolved in the water, the decomposition of organic matter, the respiration of plants and animals, the development of parasites responsible for certain diseases and the proliferation of blue-green algae which release toxins. It acts as well on the electrical conductivity as on the pH, which allows the identification of the origin of these waters and possible mixtures (AFNOR, 1985).

II.1.2. The pH :

The pH value obtained at the main collector is slightly basic, mainly due to the use of caustic soda during cleaning and washing operations at the CIP, but this value does not exceed the value authorised by the Moroccan standard for direct discharges which is 8.51.

The pH, which indicates the alkalinity of the wastewater, is crucial for the growth of microorganisms, which generally have a pH optimum of 6.5 to 7.5. When the pH is below 5 or above 8.5, the growth of microorganisms is directly affected. In addition, the pH is an important element for the interpretation of corrosion in the pipes of wastewater treatment plants.

11.1.3 Conductivity :

The measurement of conductivity allows a quick but very approximate assessment of the overall mineralisation of water and to follow its evolution (Rodier *et al.* 2009). It is a function of the total ion concentration, their mobility, their valence and the temperature. The overall mineralisation of the effluent from the main sewer remains relatively high (3027 uScni). This result can be explained by the fact that the wastewater is highly loaded with minerals originating from milk, dairy products and buttermilk, as well as the nature of the washing solutions added to the water (soda, nitric acid) at the time of the sanitation of the processes, and also originates from the brines used for the regeneration of the softeners located upstream of the cooperative.

The increase in salinity of waters has a direct effect on aquatic flora and fauna, which manifests itself in migrations and sometimes in mortality. Indeed, a very high osmotic pressure can cause, at the level of the fish gills and other external organs, diffusion phenomena through the cell walls and sometimes the death of the corresponding cells. Above 3000uscni, conditions are unfavourable for a normal ecological balance (Bremon & *al.*, 1977).

11.1.4 Suspended solids (SS) :

The TSS at the level of the main collector gives a value of 300 mg/l. This value is very high compared to the value fixed by the Moroccan standard on the limit values of direct liquid discharges, which is 50 mg/l. This is mainly due to the churning of butter and the sanitation of the installations of the production workshops, operations which produce whitish effluents.

From an ecological point of view, SS levels above 10mg/l can prevent light penetration, reduce dissolved oxygen, compromise the development of spawning grounds, reduce the available food stock and thus limit fish development by creating imbalances between the different species. Asphyxiation of fish by clogging of the gills is often the consequence of high TSS content, especially at the time of periodical emptying of dams. Deposits in hold areas can lead to anaerobic growth, with the usual consequences (Rodier *et al.*, 2009).

The abundance of suspended matter in the water reduces the luminosity and thus lowers the productivity of a watercourse, and causes a decrease in dissolved oxygen by slowing down the photosynthetic phenomena, which contribute to the aeration of the water. Similarly, they are responsible for the physical clogging of the soil, which leads to a decrease in its permeability and consequently a destruction of the quality of the soil during irrigation (Landreau A., 1987).

11.1.5 Dissolved oxygen :

Oxygen is one of the main dissolved elements. It is necessary for the metabolism of aerobic organisms and the maintenance of aquatic life. The content of this element at the level of the main collector is relatively low (2mg/l), which underlines the deoxygenation of the discharges and testifies to the abundance of oxidizable organic matter, the origin of which can only be dairy products. However, it should be noted that the quantity of dissolved oxygen, although low, remains sufficient to allow the degradation of the organic matter to begin. This degradation remains however very limited because of the short distance travelled by the effluents between the source and the sampling point.

11. 1.6 Biochemical and chemical oxygen demand (BODg and COD) :

The discharge of an industrial effluent causes a considerable increase in the organic load in the receiving environment. This results in an oxygen consumption which is expressed by the biochemical oxygen demand **(BOD$_5$)** and by the chemical oxygen demand **(COD).** These two parameters are the main indicators of organic pollution and allow an indirect measurement of the overall polluting load of the discharge.

The daily organic load noted at the level of the main collector of the cooperative is 2300 mg of O2/l for BOD5 and 3900 mg of O2/l for COD, these values are particularly high. These values are particularly high and largely exceed the value authorised by the said standard which is 100 mg/l for BOD5 and 500 mg/l for COD.

These results are explained by the fact that the collector drains effluents containing the residues of multi-products used in the different cleaning operations (CIP, production rooms, water treatment room, milk preparation room and its derivatives) using various detergents and additives as well as disinfection by sodium hypochlorites, and the use of soda (CIP station).

1.1.7 7.phosphorous materials :

Phosphorous materials exist in wastewater in various forms: -Organic: soluble or insoluble.

-Inorganics: Orthophosphates.

The predominance of total phosphorus is also very pronounced in the overall raw discharge from the main sewer, with a value of 14mg/l. This high content is not due to the effluent from the CIP neutralisation, but rather to the treatment of the plant's water supply pipes against corrosion and scaling, as well as to the active additives added to the detergents. These help to reduce the hardness of the water, thus facilitating the emulsion of oils and greases and keeping the dirt in suspension in the water; the alkalinity of the medium, modified, favours the detergent action. From an ecological point of view, phosphorus plays an important role in the development of algae; it may encourage their multiplication in reservoirs, large pipes and lake waters, where it contributes to eutrophication.

1.1.8 8 Ammonium (NH4+):

Ammoniacal nitrogen is common in water and usually reflects an incomplete degradation process of organic matter. The ammonium value obtained at the main discharge collector is high (29.5mg/l). This is mainly due to the high nitrogen composition of the milk.

Biological oxidation of ammonium can develop anaerobic zones in some parts of the distribution systems and thus cause corrosion of pipes, especially copper ones. Ammonia also has the disadvantage of requiring an increase in chlorine consumption during disinfection and of producing undesirable organochlorine compounds (mineral or organic chloramines). In practice, per milligram of nitrogen of ammoniacal origin, about 10 mg of chlorine is needed to form chloramines and decompose them into nitrogen gas (Rodier, 2009). In the aquatic environment, the permanent presence of ammoniacal nitrogen at low doses creates a certain habituation of fish to the toxicity. On the other hand, repeated exposure with short time intervals can lead to cumulative toxicity, as elimination from the body is very slow (Rodier *et al.*, 2009).

11.2 Assessment of organic pollution of wastewater in COLAIMO :

For a better appreciation of the origin of the wastewater in these effluents studied, the calculation of the COD/BOD5, BOD5/BOD5 and TSS/BOD5 ratios is of great interest (**Table 13**).

Table 13: COLAIMO wastewater ratios :

COD/BOD5	1 ,69
BOD5/COD	**0,58**
MES/DBO5	**0,13**

The use of these characterisation parameters is a good way to give a picture of the degree of pollution of the raw effluent of the cooperative and also to optimise the physico-chemical parameters of this wastewater in order to propose a suitable treatment method.

11.2.1 COD/BOD ratio5 :

The COD/BOD5 ratio allows us to deduce whether the wastewater discharged directly into the receiving environment has the characteristics of domestic wastewater (COD/BOD5 ratio less than 3). The COLAIMO wastewater has a COD/BOD5 ratio of 1.69 (**Table 13**), which is in line with the ratio of COD/BOD5 of less than 3 for predominantly domestic urban wastewater. Therefore, it can be concluded that although the wastewater from this discharge has a high organic load, it is easily biodegradable. Examination of this ratio highlights the biodegradable nature of this wastewater, for which biological treatment would appear to be entirely appropriate.

11.2.2 BOD5/COD ratio:

To characterise industrial pollution, the BOD5/COD ratio is often considered, which gives very interesting indications on the origin of wastewater pollution and its treatment possibilities. For our study, this ratio is about 0.58 (**Table 13**). This is the general case for discharges with a high organic load. This organic load makes the wastewater quite unstable, i.e. it will quickly evolve into "digested" forms with the risk of odour release. Indeed, the wastewater in this sewer is predominantly organic.

11.2.3 TSS/DBO5 ratio:

The TSS/BOD5 ratio of this wastewater is 0.13. Furthermore, the **COD/BOD5** ratio is low (1.69), which allows us to deduce that the organic matter load in the wastewater from this collector is easily biodegradable.

11.3　Results of microbiological analyses :

Requested analysis (reference of the analysis method)	Result with measurement unit
1) Enumeration of yeasts and moulds (V08-059 March 1995)	2800 CFU / ml
2) Total mesophilic load (NM ISO 4833:2008)	5400 CFU / ml
3) Total coliform count (NM ISO 4832:2008)	Absence / 1ml
4) Fecal coliform counts (V08-060 March 1996)	Absence / 1ml
5) Enumeration of staphylococcus pathogenes (NM ISO 6888-2: 2008)	Absence / 1ml
6) Testing for Salmonella in 25g (NM ISO 6579:2002/Amd 1:2008)	Absence

According to the results of the microbiological analyses of the effluents of COLAIMO, the bacterial load of these discharges is very low with the presence of yeasts and moulds and a normal total mesophilic load and the absence of pathogenic germs such as total and faecal coliforms and salmonella. These values are well above those of the domestic wastewater of the city of Oujda. The survival of these species is particularly dependent on the environmental conditions at the discharge level. The discharge from the cooperative hinders their growth and prevents them from multiplying, due to the pH and high temperature and the presence of chemicals such as disinfectants, which inhibit the proliferation of these germs, thus indicating the absence of animal faecal contamination.

Part V
The design of a Selective Biological Reactor (SBR) treatment plant for the effluent of COLAIMO

Even with the application of clean technologies, there is always a surplus of polluted water. In order to do this, it is necessary to know the proportion of pollutants in the effluent to be treated, according to the results of the analyses carried out on the effluents of COLAIMO by evaluating the ratios of the pollution of these wastewaters, the biological treatment of these effluents is conceivable insofar as the latter present a strong biodegradability and a weak toxicity.

Biological treatment processes are diverse and can be carried out using either supported or suspended microorganisms and each process has its own advantages and disadvantages as mentioned in the literature review of this report, it remains to choose the most suitable process that meets the requirements of the cooperative.

Table (14) illustrates a comparison between the different processes in terms of treatment performance.

**Table 14: Treatment performance of different processes (
International Office for Water, 2011)**

	BOD5	COD	MY	N-NK	P-Pt
Bacterial bed	60%	60%	-	50%	-
Bio filter	75%	85%	70%	-	-
Lagoon	80% a 90%	80% a 90%	70%	60%	50%
Active sludge	80%	75%	90%	70%	80%
SBR	99%	99%	98%	94%	87%

From a technical point of view, the different treatment processes generally achieve good treatment performances, but from this table we can see that the SBR system (Sequential Biological Reactor) is the most efficient and has the highest treatment efficiency.

Among the most suitable and widely used processes for food industry effluents, we distinguish ; the natural and aerated lagoon, and the active sludge, according to a study carried out to see which of these processes is the most adapted according to the nature of the effluent, and the total surface necessary, we concluded that the major problem of the cooperative is a problem of space thus it is impossible to carry out a plant by lagoon even if this one represents a less expensive financial cost, The solution is to choose a compact process that does not require a lot of space within the industrial unit and that has high purification performances. This is the case of the SBR process (Sequential Biological Reactor) which is the most suitable compact process for the effluents of the agro-food industries.

I. Sequential Biological Reactors (SBR):

The RBS system is not a recent development as is generally believed. The first RBS was installed in the early 1900s and operated on an alternative fill and drain system. The RBS system was effective in treatment but required constant monitoring. It was abandoned and resurfaced in the late 1970s with the development of reliable and inexpensive automatic control devices. The evolution and popularisation of this equipment has meant that today the RBS process has become very competitive in many respects: economy, performance and reliability.

Sequencing Batch Reactors (SBRs) are processes in which the biological reaction phases and settling processes are carried out in a single tank. These processes, which were first implemented before active sludge systems, have gained renewed attractiveness due to their recognised advantages: efficiency, modularity, adaptability, and possible automation. Moreover, the RBS process, in comparison with a continuous process, exerts a selection pressure by playing on the different growth kinetics relativising the presence of one or the other species of micro-organisms. The consequence is the minimisation of the occurrence of filamentous bacteria by applying feeding and non-feeding conditions. On the other hand, the RBS process has an advantage over continuous processes in relation to the reaction kinetics: at the end of the filling the substrate concentrations and therefore the rates are maximum.

At present, there are companies marketing different types of SBRs for a wide range of wastewater applications, especially in North America and the countries of application are in North and South America, Northern Europe and South East Asia.

The operation of an RBS consists of five basic operations during a working cycle: filling, reaction, static settling, purging (emptying the reactor), and then rest.

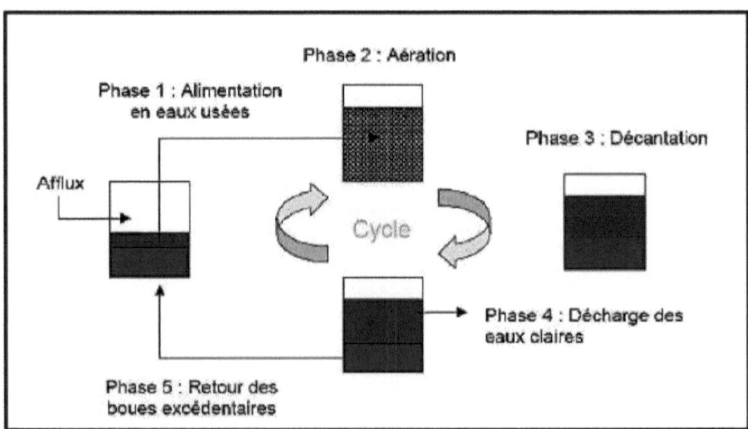

Figure 9: Various stages of operation of an SBR system

The 5 phases are carried out in 6 hours (4 times in 24H)

- Phase 1: Supply of effluent to chamber 2. (Max: 20min).

- Phase 2: Intermittent aeration© (approx. 4h).

- Phase 3: Decanting of the heaviest materials (90 min).

- Phase 4: Discharge of clear water to drain (Max: 20min).

- Phase 5: Return of excess sludge to chamber 1 (Max: 2 min).

- **<u>Phase 1: Wastewater supply</u>**

The bioreactor is supplied with wastewater by the action of compressed air. The effluent passes from chamber 1 to chamber 2. With denitrification and phosphorus release. Only water without solids is sent to the bioreactor.

- **<u>Phase 2: Aeration</u>**

During this phase air, and therefore oxygen, is introduced into the water being treated. The oxygen reactivates the bacteria, which triggers the process of decomposition of dissolved matter, oxidation of carbon, nitrification and absorption of phosphorus. The aeration is intermittent (cycle).

- **<u>Phase 3: Decanting</u>**

During this phase there is no oxygen supply, so chamber 2 is at rest. By the action of sedimentation, the active sludge settles to the bottom. The clear water rises to the surface and the remaining impurities settle to the bottom of the tank.

- **<u>Phase 4: Clear water discharge</u>**

During this phase, the clear water resulting from the biological process is discharged outside the plant.

- **<u>Phase 5: Return of surplus sludge</u>**

The active sludge at the bottom of chamber 2 is returned to chamber 1. At the end of this phase, the cycle restarts at phase 1. The sludge in chamber 1 must be removed once a year or when chamber 1 is 70% full.

The pond is equipped with a level control system associated with an automatic controller that orchestrates the different sequences. The supply is necessarily made from a lift station with a pump flow sufficient to absorb the pollutant flow resulting from a storm. The construction of a pollution basin is an additional advantage. The effluent is admitted during the so-called aeration phase. When the high level is reached, the supply is stopped and a treatment phase begins. After settling, the draining phase of the effluent to the natural environment is carried out to low level, followed by the extraction of the excess sludge to a storage silo. A new cycle is activated with the authorization of feeding. Oxygenation of the sludge is provided by a blower or booster which is also capable of powering an air lift for effluent discharge installed in place of a float pump. Excess sludge is automatically removed from the tank at the end of the supernatant discharge periods (three times a day) by an extraction pump.

II. The design of an SBR treatment plant for COLAIMO's effluent:

In order to build a wastewater treatment plant, it is first necessary to design the plant and then to dimension the treatment works for a good management and operation of the built plant. In this context, this project aims at designing a wastewater treatment plant by SBR for the effluents of the Cooperative Laitiere du Maroc Oriental.

■ **Basic data :**

Nature of the sewerage system	**Separate**
Connected industry	Dairy
Daily volume	600 m3/d
Average hourly flow rate	30 m3/h
Peak hourly flow rate	60 m3/h

N.B.: As we do not have a flow measurement device, the values of flow and MVS have been taken from a report made by Mr. Jean Pierre VETEAU (expert engineer) of a water rationalisation project of COLAIMO in August 2010.

1. Pre-treatment works:

Pre-treatment consists of three main stages that remove from the water the elements that generate the following stages of treatment. Not all wastewater treatment plants are equipped with three, only screening is generalized, the others are grit removal and degreasing.

1.1 . Degritting :

The strainer is one of the first elements to be found at the entrance of a wastewater treatment plant, usually located upstream of the pumps. It allows to extract from the water, the big wastes, such as: leaves of trees, metallic objects... etc. It ensures the protection of the electromechanical equipment and reduces the risk of clogging of the pipes installed in the treatment plant.

<u>**Screening**</u>

- **<u>Design and sizing of the screen :</u>**

The performance of a screen (manual or self-cleaning) is characterised by its spacing between bars.

The spacing of the bars of the screen is determined by the choice of the size and nature of the objects accepted by the station. A compromise is also sought between the spacing of the bars and the quantity of waste to be removed (frequent cleaning of the grid).

- **<u>Grid sizing :</u>**

The vertical cross-section of the grid which also represents the cross-section of the channel is :

$$S_v = \frac{Q}{V\,a\,C}$$

With - Q: design flow m3/s

- V: the velocity of passage through the grate, which should be between 0.6 and 1.2 m/s (greater than 0.6 m/s to avoid sand deposits and less than 1.2 m/s to avoid high head losses)

- a: free passage coefficient $a = \dfrac{e}{e+s}$ (e: spacing between bars and s : Width of a bar)

- C: clogging coefficient which is between 0.4 and 0.5 for the mechanical screen and between 0.1 and 0.3 for the manual screen. (The screen used will be a mechanical screen so we will use a clogging coefficient between 0.4 and 0.5).

> Let C= 0.45, Q=0.00694 m3/s, e=40 mm, s=10mm, a=0.8 and V=0.9m/s.

On the other hand we fix l= 1m. The grid will be composed -with the given dimensions- of a number of bars equal to N, such that :

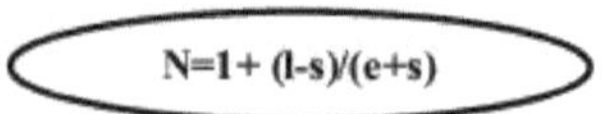

- The number of bars in the grid will therefore be **21 bars.**

- The width of the grid is $L = n \times s + (n+1) \times e$ → **1.09 m**

Finally the dimensions of the grid

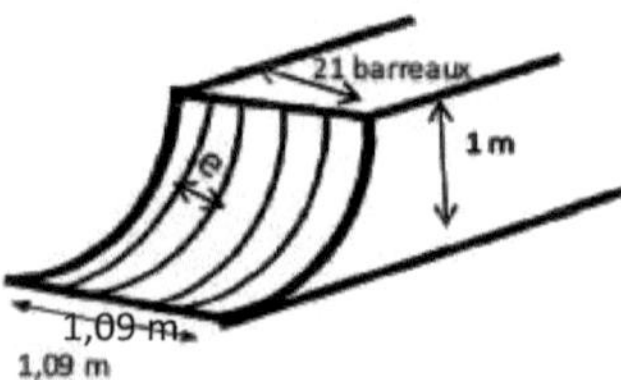

Finalement les dimensions de la grille :

l= 1m   ;   L=1 .09m

e=40mm   s=10mm

Sv= 0,0214 m²

Nombre des barreaux : 21 Barreaux

Number of bars: 21 bars

Follow-up :

-Daily visual inspection of electromechanical equipment

-Daily inspection of the quantity of waste retained.

- **Waste disposal :**

-Storage in landfill.

-Incineration after compaction.

-Prohibit their disposal in agriculture.

1.2 Raw water collection : Lift station

With this configuration, any gravity supply is impossible. The effluents directed in treatment will be obligatorily by the intermediary of a lifting station

The construction of sewage systems and wastewater treatment plants often requires the installation of sewage lifting pumps due to the large differences in level.

The simplest and safest installation is to place one or more submersible pumps directly into the sump or pumping chamber. The motors, bearings and electrical connections are enclosed in a hermetically sealed enclosure, which protects them from water and shock.

The design of this equipment facilitates maintenance and repairs by the simplicity of replacing all its parts. The connection device is automatic, the installation and removal of the pump is done, without intervention in the station, by simple unlocking. The pump is lifted by means of a jib crane equipped with a manual chain hoist.

The cost of excavation and installation is kept to a minimum, as the volume of the post is practically the same as the volume actually required, and the sheeting is manufactured using industrially prefabricated reinforced polyester panels, with or without a cover.

1.3 **Screening :**

The screen allows for even more precise pre-treatment by retaining fibrous particles (e.g. hair) and finer elements (500um to 3mm). After passing through the distribution box, the wastewater to be filtered arrives through an overflow on the upper quarter of the filter drum. The water passes through the screen while the solids are stopped, carried along by the rotation of the cylinder and removed by a scraper on the opposite side to the feed.

A spring device ensures permanent contact between the scraper and the drum, automatically compensating for wear.

At the bottom, the water flows through the drum in the opposite direction. Thanks to the kinetic energy acquired in its fall, it ensures the cleaning of the cylinder, eliminating the particles that escaped the scraper. The rotary screen used will be adapted to the flow rate of about 3 m3/h and accompanied by a waste compactor.

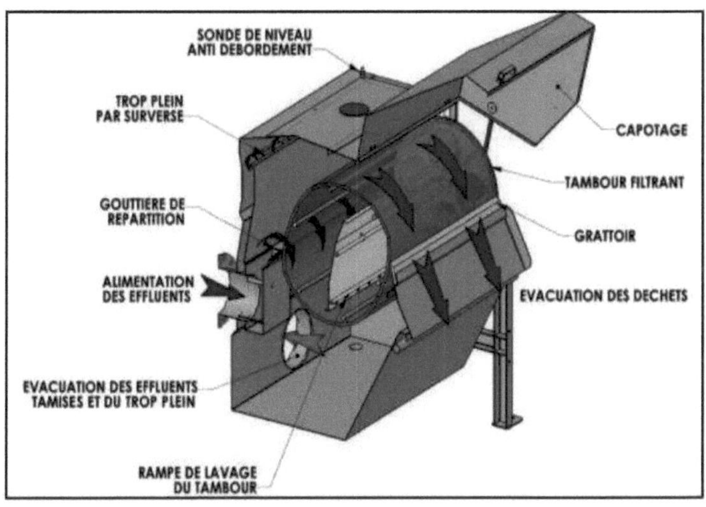

Figure 10: Diagram of a rotary sieve

The technical characteristics of the sieve (type, mesh, length of the sieve and the speed of the drum ...) will be given by the company in charge of the installation of the treatment plant.

1.4 Float :

We can remove a significant proportion of the impurities in wastewater by causing it to rise to the surface and skimming it off.

This is particularly the case for fats and oils of lower density than water. The operating principle of the grease separator is based on a simple physical law: the difference in density. In order to accelerate the rise of the fatty particles, the effluent will be emulsified by air blowing, which reduces the time of passage in this structure and prevents, by the stirring it causes, any sedimentation of heavy matter. The structure is compartmentalised in such a way that the grease collects on the surface in a quiet zone, allowing automatic extraction, while the water and sludge are evacuated.

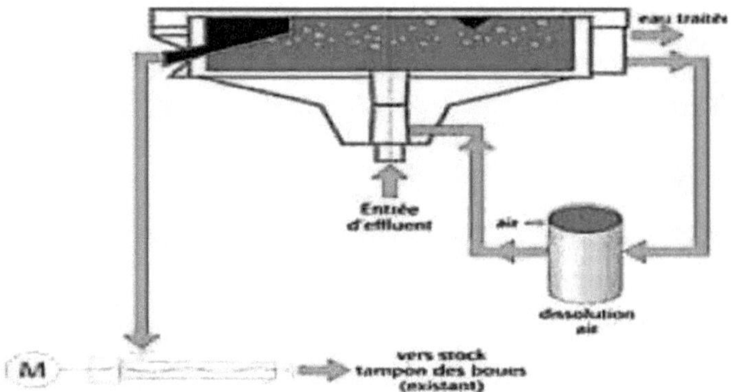

Figure 11: Principle of flotation

2. Neutralization :

Industrial waste contains alkaline or basic materials that require neutralisation, in our case the pH of the plant is basic but it can become acidic depending on the quantity and nature of the chemical products added to avoid this problem we proceed to a neutralisation before the passage of these waters to the biological reactor, because the pH plays an important role in the bacterial activity, so the pH must have a value close to the calco-carbonic balance to protect the materials against corrosion and against scaling.

The effluent of the cooperative has a basic pH. We propose a neutralisation with carbon dioxide, at the outlet of the flotation device, in the treatment basin. The natural product CO_2 is non-toxic, non-flammable, safe to store and easy to use. We can then rightly say that CO_2 allows for the most ecological neutralisation of alkaline waste water, and to minimise the risks we propose neutralisation with lime if the pH of the water becomes acidic. A pH measurement probe will be installed in this tank. A dosing pump, controlled by the pH measurement in the tank, injects CO2 or lime to correct the pH.

1. **The Reactor: Treatment tank**

■ **Volume :**

The sizing of the reactor volume is based on the mass loading which depends on the desired reactor performance and the concentration of solids in the reactor.

$$\text{V nécessaire} = \frac{DBO_5 \times Q \times 10^{-3}}{Cm \times MVS}$$

With :

BOD5 : Biochemical oxygen demand for 5 days

Q : Flow in m3/d

Cm: Mass loading in kgDBO5/kgMVS.d

MVS: Biomass concentration in the reactor in g/L.

That is, BOD5=2300mg/l, Q=600 m3/d, Cm=0.09 Kg BOD5/kg MVS.d and MVS=2.9g/l.

The volume required for the reactor is:

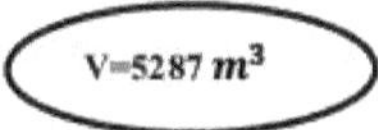

There are two possible solutions:

Construction of a single reactor of 5287 m^3 :

A single reactor presents a risk of untreated effluent being discharged in the event of failure. Furthermore, the size of the system would not fit into the landscape for such a basin.

Construction of 2 reactors of 2644 m^3 each:

This solution would put in place 2 reactors, in addition to being more "discrete" in the landscape, in case of failure of one of the reactors the effluents could still be treated by the 2nd reactor.

Ventilation :

Aeration is provided by a fine bubble aerator with clock control. Increasingly reliable and economical fine bubble systems are available. Fine bubble aeration is characterised by its high efficiency and energy performance, and is twice as energy efficient as surface aeration systems. The presence of an agitator that is controlled by the aerator's non-operation is essential to optimise the treatment.

- Oxygen requirement :

$$M_{O_2} = a_0 Q(S_e - S) + bXV + \delta c Q C_N$$

a_D : Theoretical rate of combustion of O_2 to degrade one mass unit of substrate in g 0_2 I g BOD_S

b: Consumption rate of 0_2 for endogenous respiration ел g O_2 / g bones / unit of time

c:Consumption rate of O_r for nitrification in dO_2 / g nitrogen

3: Та их of nitrification (function of C_{hl} and т)ь

C_N : Nitrogen concentration of water a trailer

Total daily oxygen requirement is: **2299 KgO2/d**

Anoxia :

The nitrogen having been nitrified in the aeration tank, its denitrification is planned in the same tank, by stopping the aeration. The anoxia cycles will be specified at start-up and optimised during this period. These cycles will follow the aeration periods and precede the settling periods. The oxygen contained in the nitrates is used to oxidise some of the carbon pollution. To achieve this result it is necessary to provide a vigorously agitated, but not aerated, tank.

■ **Phosphorus treatment :**

Phosphorus removal can be achieved by a physico-chemical treatment based on the ability of ferric chloride *Fe Cl₃* to combine with phosphate ions to form a precipitate of iron phosphate *FePO₄* , a salt which is not very soluble in water and which precipitates in a colloidal state, which is known to be difficult to decant. This precipitate is absorbed by an excess of metal hydroxide (OH)₃ . This product will be added to the reactor by means of a simple dosing pump.

■ **Decanting :**

As soon as the aerator stops, the 2nd operation of the cycle begins. The treatment tank becomes a very large settling tank. The sludge separates from the water by sedimentation and settles to the bottom of the tank.

■ **Water drainage :**

After a sufficiently long settling time, the third and final operation of the cycle can begin. The treated water is returned to the surface by means of a floating device, controlled by a 24 hour clock. A level regulator ensures that this device stops automatically when all the accumulated effluent has been discharged before the end of the time allowed for this operation. A new operating cycle is automatically started as soon as the "Evacuation" period is over. The water is sent to the measuring channel.

4. Sludge extraction :

Periodically, after the aerator has been switched off and the tank has been left to rest, the sludge is pumped back to the lifting station for mixing.

There is no sludge recirculation as such, but some of the decanted sludge is retained within the reactor at the end of the cycle. This is used to inoculate the wastewater brought in for the next cycle.

5. Self-monitoring and automatic sampling:

It is useful to install a flow measurement channel at the inlet and outlet of the station to check that it is working properly.

A Venturi flume which will be composed of a flow meter and automatic sampler to perform self-monitoring. The same type of equipment will be installed upstream in order to carry out self-controls on the raw water. The flow measurement will be in open channel by an ultrasonic probe.

6. Valorisation of milked effluents:

After treatment of the slag effluent it is either discharged to the receiving environment or reused within the plant.

Industrial reuse increases plant productivity by saving water resources and reducing the volume of wastewater.

- The cooperative's sewage effluent can be used for :

watering of green areas instead of using treated water from the
water treatment plant,
cleaning of equipment and premises,
Cooling water for air conditioning,
The supply of individual toilet flushing circuits.
Wastewater reuse projects contribute to integrated water management
They are particularly strategic in arid and semi-arid countries where the pressure on water resources is high and where there is competition between different water uses in a context of climate change.

7. Expected treatment performance:

Parameters	Contents	Performance
BOD5	23 mg O $/l_2$	99%
COD	39 mg O_2 /l	99%
MY	6 mg/l	98%
PT	1.82 mg/l	87%

8. Diagram of the Sequential Biological Reactor (SBR) treatment plant planned for the effluent of COLAIMO:

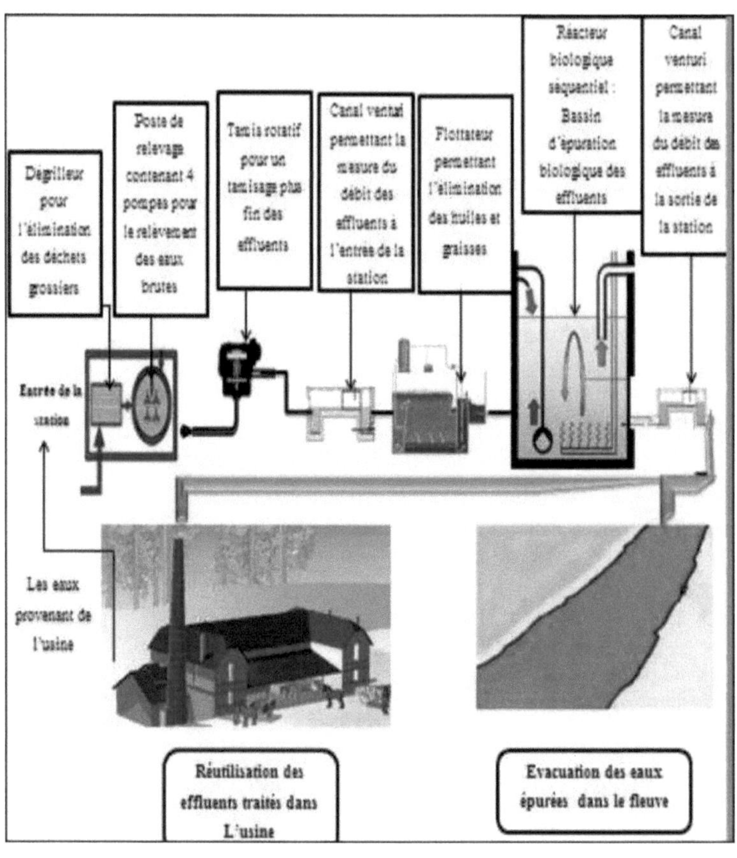

Dégrilleur pour l'élimination des déchets grossiers
Poste de relevage contenant 4 pompes pour le relèvement des eaux brutes
Tamis rotatif pour un tamisage plus fin des effluents
Canal venturi permettant la mesure du débit des effluents à l'entrée de la station
Flottateur permettant l'élimination des huiles et graisses
Réacteur biologique séquentiel : Bassin d'épuration biologique des effluents
Canal venturi permettant la mesure du débit des effluents à la sortie de la station
Entrée de la station
Les eaux provenant de l'usine
Réutilisation des effluents traités dans L'usine
Evacuation des eaux épurées dans le fleuve

Conclusion

The main objective of this work is the characterization of the wastewater of the Cooperative Laitiere du Maroc Oriental (COLAIMO) in view of the implementation of a compact process allowing the purification and the valorization of these dairy effluents for an integrated management of the water resources and for the preservation of the environment

Through the results of the physico-chemical and bacteriological analyses carried out, the wastewater of the COLAIMO presents values of the major physico-chemical parameters of pollution which relatively exceed the general limit values of the direct and indirect discharges in the receiving environment, which represents a risk of environmental pollution for the latter, hence the need for a treatment of this raw wastewater.

At the end of the evaluation of the degree of organic pollution, it can be seen that all the parameters studied (in particular with BOD5, COD and TSS) place the wastewater analysed in the medium to high concentration range.

The evaluation of the pollution ratios of this wastewater shows that it is easily biodegradable and the COD/BOD5= 1.69 value underlines the biodegradable character of this water to which a biological treatment seems quite suitable.

The treatment of this wastewater is necessary to produce an effluent that respects the standards of direct and indirect discharges according to the Moroccan Ministry of the Environment.

A comparative study of the different biological purification processes shows that the Sequential Biological Reactor (SBR) process is the most suitable for the purification of COLAIMO's effluents because it is the most compact and the most efficient. The installation of this system within the cooperative will allow a better management and rationalisation of COLAIMO's water, as well as the preservation of the environment for a sustainable development

Bibliographic references

ABHM (2010), Agence du Bassin Hydraulique de la Moulouya. Etude Du Plan Directeur D'amenagement Integra Des Ressources En Eau Du Bassin De La Moulouya (PDAIRE), Mission II : Developpement des ressources en eau du bassin, Sous Mission II.2: Bilan besoins-ressources 2030, préservation de l'environnement Et Optimisation Technico-économique Mars 2010, pp. 23-112.

AFNOR, Recueil des normes franchises des eaux, methodes d'essais, AFNOR, 2nd edition, PARIS; 1985, p. 566-622.

Beauchamp (J.) - Coastal pollution. D.E.S.S. Qualite et Gestion de l'Eau, Univ. Picardie Jules Verne 2003.

BENSASSI A., JAOUAD A., MANDI L. & BOUSSAID A., Comparaison du traitement des margines par les levures, First Int. Symp. Management Liquid Solid Residues (MALISORE), Mohammedia, Morocco, April 2004.

BENYAKHLEF M., NAJI S, BELGHYTI D., Caracterisation des rejets liquides d'une conserveie de poissons, Bull. Soc. Pharm. Bordeaux, 2007, 146, 225-234, p. 231.

BENYAKHLEF M., Etude et caracterisation des rejets des industries agroalimentaires en vue d'une economie et une optimisation des circuits des eaux (Region d'Agadir), PhD thesis in sciences, Faculte des sciences de Kenitra, 2008, p. 138.

BREMON R., PERRODON C., Parametres de la qualite des eaux, Rapport du Ministere de l'Environnement et le Cadre de Vie, direction de la prévention des pollutions services des problemes de l'eau, Paris France, 1977, p. 258.

5.　**Castillo de Campins (2005),** Study of a compact process for aerobic biological treatment of dairy effluents. D. in Ecological, Veterinary, Agronomic and Bioengineering Sciences, INSA, Toulouse- France, 2005.

COLAIMO (2009) Rationalisation and management of water consumption in the industrial sector in Morocco:2-17

National Environment Council, Economic Instruments for Environmental Protection in Morocco, 2009, p. 8.

Corthondo. Trepos F. (2004) Treatment of dairy effluents: 1-13

Donkin M.J. (1997). Bulking in anaerobic biological systems treating dairy processing wastewater.Int.J.Dairy Technol. 50:67-72

F.Pettillot (1974) Prevention and control of pollution and nuisances in cheese dairies. Service de l'environnement industriel, December.

Garrido J.M.,Omil F.,Mendez R.,Lema J.M (2001). Carbon and nitrogen removal from a wastewater of an industrial dairy laboratory with a coupled anaerobic filter sequencing batch reactor systems,Wat. Sci. Technol.43:249-256.

HAMADANI A., CHENNAOUI M., ASSOBHEI O. & MUNTADAR M., 2004: Caracterisation et traitement par coagulation-decantation d'un effluent de laiterie, INRA, EDP Sciences, DOI : 10.1051/lait : 84, p. 317.

Khoudir A., Lamari H., Louchli S. (1997) Fixed bed biological treatment of dairy wastewater. Eau Ind.Nuisances 203 :37-39.

LANDREAU A., La reutilisation des eaux usures epurees par le sol et le sous-sol :adequation entre la qualite de l'eau, l'usage et la protection du milieu naturel, Seminaire sur les eaux usees et le milieu recepteur, Casablanca 9-11 avril 1987, p 1-13.

Ministry of the Environment of Morocco, (2002). "Normes marocaines, Bulletin officiel du Maroc", N° 5062 du 30 ramadan 1423. Rabat.

Moletta R., Torrijos M. (1999) Environmental impact of dairy effluents, Tech.Ing, F-1500 :1-9.

International Office for Water (2011). The ban on the disposal of by-products of pollution (European Regulation) problems and solutions.

ONEP, Documentary base on the management of industrial discharges
April 2007, p34.

UNEP. (2000) Cleaner production assessment in dairy processing. United Nations Publications: 95p

R. Hamdaoui. "Physico-chemical characterization and treatment of wastewater from SETEXAME. Kenitra ". Memoire DESA. 2006. Fac. SCI. Kenitra. Universite Ibn Tofail. 100p.

RODIER J., LEGUBE B., MERLET N., et al: L'analyse de l'eau, 9eme edition, DUNOD Paris, 2009 p. 1217, 1381.

SPREER E. (1991). L;iclologi;i Industrial. Editorial Acribia.

WHO (World Health Organization. Division of Environmental Health United Nations Environment Programme Global Environment Monitoring System, Global pollution and health results of related environmental monitoring, Global Environmental Monitoring system, UNEP, 1987, p. 12.

yes I want morebooks!

Buy your books fast and straightforward online - at one of world's fastest growing online book stores! Environmentally sound due to Print-on-Demand technologies.

Buy your books online at
www.morebooks.shop

Kaufen Sie Ihre Bücher schnell und unkompliziert online – auf einer der am schnellsten wachsenden Buchhandelsplattformen weltweit! Dank Print-On-Demand umwelt- und ressourcenschonend produzi ert.

Bücher schneller online kaufen
www.morebooks.shop

info@omniscriptum.com
www.omniscriptum.com